NOCTURNAL ANIMALS

**Recent Titles in
Greenwood Guides to the Animal World**

Flightless Birds
Clive Roots

Hibernation
Clive Roots

NOCTURNAL ANIMALS

■ Clive Roots

Greenwood Guides to the Animal World

GREENWOOD PRESS
Westport, Connecticut • London

Library of Congress Cataloging-in-Publication Data

Roots, Clive, 1935–
    Nocturnal animals / Clive Roots.
        p.   cm.—(Greenwood guides to the animal world, ISSN 1559–5617)
    Includes bibliographical references and index.
    ISBN 0–313–33546–X (alk. paper)
    1. Nocturnal animals.   I. Title.   II. Series.
    QL755.5.R66 2006
    591.5′18—dc22   2006010622

British Library Cataloguing in Publication Data is available.

Library of Congress Catalog Card Number: 2006010622
ISBN: 0–313–33546–X
ISSN: 1559–5617

First published in 2006

Greenwood Press, 88 Post Road West, Westport, CT 06881
An imprint of Greenwood Publishing Group, Inc.
www.greenwood.com

Printed in the United States of America

∞™

The paper used in this book complies with the
Permanent Paper Standard issued by the National
Information Standards Organization (Z39.48–1984).

10 9 8 7 6 5 4 3 2 1

For Jean
For the many years of love, companionship,
and shared concern for animals,
from London Zoo to Vancouver Island

Contents

Preface

Finding food and mates, establishing and defending a territory, and avoiding predators and inclement weather are the daily functions of wild vertebrate animals. To behave in the same manner in the dark needed considerable improvement to their already impressive adaptations for survival, yet this was achieved by almost half of all living species.

A barn owl can locate a mouse in total darkness solely by the tiny sounds it makes as it chews grass seeds. The owl's large, dished face funnels these sounds to the huge ear openings beneath its feathers, which are uneven; this improves sound reception from both the vertical and horizontal planes, allowing pinpoint accuracy even when vision is compromised. On a moonless night in the Mohave Desert a rattlesnake's flicking forked tongue picks up scent particles left by a kangaroo rat, and transfers them to organs in the roof of its mouth, from where a message is flashed to its brain that directs the snake along the rodent's trail. The rattlesnake also has a sixth sense, a heat-sensing system that includes deep facial pits containing nerves and receptor organs that detect fractional temperature variations and can locate the rodent by its radiant warmth.

The tiny pipistrelle bat, weighing just a fraction of an ounce, is the ultimate example of natural miniaturization, a product of many years' refinement of its senses and adaptations. Flying fast at night over the English countryside, it emits ultrasonic sounds that rebound off the trees and its insect prey. Its sense of hearing is so acute that it can act upon the returning echos in time to avoid obstacles or scoop up an insect with its wing.

This book is an introduction to these and many other night creatures whose highly developed senses and unique behavior are unequalled among animals.

Introduction

Twice daily, as the sun rises and later sets, animals around the world change places, taking turns to share the facilities and food supplies of a particular niche. Only migration to a warmer climate or the seasonal inactivity of hibernation affect this routine. By reversing what is usually considered normal behavior and being active after dark, the nocturnal animals make use of the habitats and niches previously occupied by others, sometimes just minutes before, without becoming competitors. This is the original nightshift, the animals that use the environment's resources vacated by the day group, and the only difference is the lack of sunlight, which nocturnal animals do not need. Life in the dark is so enticing that its occupants comprise almost half of all living vertebrates.

Throughout the world, in deserts, grasslands, and forests, and in the tropics and cooler regions, most niches support two separate populations. As the insect-hawking swallows and martins leave the California skies at dusk, their place is taken by nightjars and poorwills. In central Nebraska the daytime hunting grounds of the red-tailed hawk become the nighttime domain of the great horned owl, and near the Mexican border a gila monster crawls from its hole onto a rock still warm from the sun, where a chuckwalla basked earlier. Blue and gold macaws in Trinidad's Nariva Swamp tear a last oily fruit from a palm tree before flying off to roost, and when it is quite dark the echolocating oil birds leave their caves and fly to those same palm trees, whose fruit is also their favorite. Most of these night creatures are not familiar animals. When darkness falls and human activities continue in a world of artificial light, the animals active outside are seldom seen or heard; the habits and behavior of many night creatures are difficult to study and are poorly known.

Like the daylight or diurnal animals, which are not necessarily active all day, nocturnal species may not be active all night. Although in general they regulate their waking by the setting of the sun and their retiring by sunrise, there is a great deal of variation in between. Twilight witnesses the activity of a very specific fauna, known as the crepuscular animals, which are not seen during the darker hours. Others vary their activity throughout the night with alternating periods of resting and eating. Few truly nocturnal animals in which sight is the dominant sense are active during the early hours of the morning when light levels are at their lowest, because even species with highly developed night vision cannot see in almost total darkness, and hearing and touch are then more important. Whereas some night animals, such as the northern insectivorous bats, are strictly nocturnal and very light-sensitive and are rarely seen flying during the day, other nocturnal creatures may appear on dull days or even in bright sunlight to bask briefly or to hunt if they were unsuccessful the previous night. Others, generally small species that have a short rhythm of activity, may divide each twenty-four hours into periods of activity and rest, and these are known as arrhythmic animals.

The advantages of nocturnal activity have been overrated, however. Nocturnal habits do not necessarily confer a more secure lifestyle, and where safety from predation is concerned life is no easier for the nocturnal animals than for those active by day. Constant alertness and the avoidance of predators is the main preoccupation of wild animals, even of most predators themselves, whether nocturnal or diurnal. All animals have enemies, most of them all the time, the fortunate few just at some stage of their lives. To ensure their continued existence they must always be prepared to take flight or defend themselves and their young. The abundance of fossilized remains proves that many could not compete in the demanding world of nature during the course of evolution. Among the survivors the mechanisms concerned with the capture of prey or the escape from predators have reached the stage where improvement seems impossible or unecessary, but this can never be the case, for any slight improvement in a predator's abilities must be countered by a similar one in its prey's defenses, and vice versa.

The nocturnal prey animals must be alert for predators at night and must find a secure hiding place for the daylight hours; they have evolved adaptations and behavior to assist their survival in darkness and in daylight. Their adaptations to ensure survival in minimum light have few equals in the animal kingdom, and at least one of the senses—sight, hearing, touch, taste, or smell—is highly developed. The echolocation powers of the bats; the highly evolved sensory cells of some snakes, which can detect minute changes in temperature; and the incredible hearing of the owls all attest to the amazing ability of animals to adapt to a variety of circumstances. When the sun rises most nocturnal animals are back in their daytime resting places, which must protect them from the sun and from predators. Back in its burrow after a night of searching for seeds, the kangaroo rat plugs the entrance with grass and soil to keep predators out. The bats have returned to their cool and moist caves until sunset, and owls are safe in tree cavities or in dense vegetation where their drab coloration blends with the leaves and shadows. The moist-skinned amphibians must find shade and moist places, and many burrow into the soil, but even the microclimate on the underside of a large leaf is adequate for some.

Animals that are food for the nocturnal predators are not necessarily nocturnal themselves. In fact, most of the prey species of the large cats are diurnal, yet make no attempt to hide while resting at night. On the open African savannah and in the shrub-covered bushveldt, impala, eland, giraffe, and zebra are fair game after dark for lions, hyenas, and leopards.

The world of darkness is a truly different world, mysterious and fascinating, where humans are at a distinct disadvantage. Darkness is the domain of the venomous snake that can follow an animal's scent trail in the dark by tasting the air with its tongue, and can register the heat radiating from its prey to a fraction of one degree. It is the domain of the owl, whose hearing is so acute it can catch a mouse in total darkness, and of the bats, the only "winged mammals" that use reflected sound waves to avoid obstacles while flying at speed.

This is an introduction to the animals that inhabit that world, how they live in the dark, and how they avoid their daytime enemies and survive until night falls again.

1 Darkness and Life

The ability to survive in darkness depends upon a combination of structural modifications and the extreme development of the senses, usually to the point where our own receptors seem feeble by comparison. Animals cannot exist in the highly competitive natural world, whether by day or night, unless their senses are highly attuned, and the acuteness of a particular sense depends upon the degree of reliance placed on it. Nocturnal activity certainly does not infer a safe way of life, and like their daytime counterparts, the night-active animals must be ever alert to the opportunities for finding food and escaping threats. The predator must eat and most of the prey must escape, with the same constant upgrading to counter the improvements made by the other, and the fossil record is rich with examples of those that could not compete.

The adaptations that have evolved to ensure survival in the dark have few equals in the animal kingdom. In addition to the development of at least one of their senses to a high degree, many animals have additional sensory systems. Some snakes have a heat-sensing mechanism known as the sixth sense, the frog's whole skin is sensitive to light, and the aquatic salamanders have organs that are sensitive to movement in the water. Most animals also have a more general sensory system in which tiny sense organs in the skin are aware of environmental changes. Many are sensitive to barometric pressure, infrasound, wind speed and direction, and can sense an oncoming storm or the effects of a distant earthquake[1] and change their behavior accordingly. Sense organs are only concerned with the reception of a stimulus, however, not its interpretation. They all have a similar basic plan, with receptor cells receiving stimuli and transmitting nerve impulses to the brain, where they are interpreted and a decision made on the most appropriate action. In addition, most animals rely on automatic responses such as reflexes, which are the involuntary reactions to stimuli, and instinct—which is the natural and largely hereditary aptitude of species to respond to environmental stimuli.

Animals rely on their senses for information on their surroundings. Sound, light, smell, taste, touch, and movement are all sources of this information if the senses can perceive it and relay it to the brain. The importance of a sense to an animal varies according to the species and its way of life or evolved adaptations, and the brain combines the messages it receives from the various sense receptors, a process called perception.

Generally, the vertebrate animal's body is well endowed with sensory receptors, which are specialized cells within a sense organ (such as the olfactory organ for smell) that monitor the environment by responding to selected stimuli and transmitting the information to the brain. The sensory organs are of several types, including somatic sensory organs, which provide information about changes to the external environment and affect the body as opposed to the mind; and visceral sensory organs, which provide information about an organism's internal environment. There are also general sensory organs, which are distributed over the surface and insides of the body; and special sensory organs, such as the lateral line of fish and aquatic amphibians, the labial heat-sensing pits of the pythons, and the scent-tasting vomeronasal organs of many vertebrates.

The natural complement to nocturnal activity for most animals is concealment during the day from the sun and from predators, and most hide so successfully that they are seldom seen. Natural hiding places include hollow trees, caves, crevices, and dense vegetation; and nocturnal animals also use artificial sites, such as attics, mine shafts, bridges, and culverts. Many burrow into the soil, either making complex tunnel systems or just letting the loose soil cover them. But these hiding places are not always safe from predators whose own adaptations aid their search for sleeping prey.

■ THE SENSES

Vision

Sight is a most important sense, surprisingly, for many animals of the dark, and special adaptations to the eye's structure allow them to cope in darkness. The animals included in this book, from the amphibians to the mammals, all have eyes of basically the same structure, with the ability to form distinct images on the retina and to distinguish changes in light intensity. However, nocturnal habits place restrictions on the use of vision in general, and although birds depend mainly upon sight,[2] many mammals place more reliance upon the senses of hearing and smell. When the other senses are well developed the lack of good eyesight is not a hindrance, as the means of sense reception is not the important factor. It is the message that is significant, by whatever means it reaches the brain, where it is deciphered to build a mental image.

Eyes must be able to receive sufficient light to have survival value at night and they are enormous in the nocturnal animals that depend mainly upon vision. They reach their maximum size in the tarsier, a small, insectivorous primate that has the largest eyes in relation to its size of all the mammals, and in the owls, which hunt close to the ground because they are shortsighted. In fact, in most nocturnal hunters night vision is myopic or short range, as they rely upon this sense only when they are within striking distance of their prey. The amphibians generally have good eyesight, and it is especially acute, with good depth perception, in the tree frogs, which leap from branch to branch in the dark. The nocturnal snakes also have good night vision, generally monocular, but with binocular vision in the tree snakes where, again, greater accuracy is required to seize a bird or bat flying past their ambush site. The nocturnal birds all have good night vision with the exception of the kiwi, which relies more on its sense of smell.

Large eyes, with increased retinal surface, larger lens, and wider pupil, are all adaptations to collect more light. The owl's increase in eye size has been achieved by developing tubular eyes, so large that they occupy half of the skull. The avian retina, unlike that of the mammals, has no blood vessels and is said to be avascular, which prevents shadow and light scattering. This is because of the presence of the pecten, a vascular structure that projects from the retina into the vitreous humor from which nutrients and oxygen diffuse. The owls have a smaller pecten than diurnal birds and therefore their power of accommodation or automatic adjustment is limited. The front-facing eyes of the predators provide binocular or stereoscopic vision—overlapping fields of sight of each eye—which increases the perception of depth and improves their ability to locate and seize their prey.

Eagle Owl *Facing forward and providing stereoscopic vision, the eagle owl's eyes are the largest of all birds in relation to its size. Tubular in shape and with a large lens and wide pupil, they occupy as much room in the skull as the bird's brain.*

Photo: Ferenc Cegledi, Shutterstock.com

Like all land vertebrates the nocturnal animal's retina contains photoreceptor nerve cells called rods and cones, which translate light into nerve impulses; the number of each type depends upon the species and its habits. Physical and chemical changes occur in these cells when light falls upon them, triggering responses that flash to the brain. Cones are of three types—red, green, and blue—and perceive and discriminate color at very high levels of light intensity. The denser the cone cells the sharper the vision in daylight, and the diurnal animals with their preponderance of cones have good daylight sight, color vision in varying degrees, but they do not see too well in the dark. The cone cells are mostly concentrated in the yellowish central area of the retina, opposite the pupil, called the macula, which contains in its center the fovea centralis. In contrast, nocturnal animals have few, if any, cones, but lots of rods, which are effective at low light levels and provide good night vision; however, they are color blind.

The fovea centralis is the tiny but very sensitive area of the retina that provides the clearest vision and greatest focus, and the blood vessels and nerve fibers go around it so that nothing interrupts the light's path to its photoreceptors. In diurnal mammals the greatest concentration of cones occurs in the fovea, which in forest species is circular to allow for the contrasting elements of their environment; grassland mammals have an elongated fovea, which allows wider focus. In nocturnal mammals the fovea has a 10:1 ratio of rods to cones.

Although rods are far more sensitive than cones, their picture is not as sharp because many rods send their messages along a single nerve, whereas each of the more numerous, but less sensitive cones, has its own nerve fiber going to the brain. Multiplying in the rods during darkness, the retina's light-sensitive pigment (called rhodopsin or visual purple), which is manufactured by combining Vitamin A with protein in the presence of darkness, absorbs most visible color and converts the light into energy. Rhodopsin does not have equal intensity to all colors, however; it is most responsive to green and falls off toward blue and red at the ends of the spectrum. As nocturnal animals cannot perceive red, or very much blue, these colors have been used to light reversed-activity displays or nocturnal houses in zoological gardens, allowing visitors to see active animals that believe they are in the dark.

Generally, nocturnal animals' retinas are packed with rods, like the owls, which have about 125 million in each eye. Some nocturnal animals also have a few cones in their retinas; others, like the echolocating bats, some snakes, and lizards, have none at all. In contrast, diurnal birds have many cones and few, if any, rod cells; they are therefore virtually helpless in the dark and flounder about if they are disturbed when roosting.

The owl's elongated tubular eyes also have some cones, giving it surprisingly good daylight vision, but not enough for color perception, so owls live in a gray world, day and night, as the pictures transmitted to their brains by the rods are in shades of black and white only. Their drab plumage, and that of other nocturnal birds such as nightjars and frogmouths, is a good indication that they do not need brightly colored plumage to attract a mate.

The formation of the light-perceiving rod and cone nerve cells varies according to the species and its habits. Many nocturnal animals can see fairly well in daylight, but their eyes must be protected from bright light. The nocturnal eye is generally larger, with a wider pupil and greater retinal surface to collect more ambient light, and because their eyes are so highly adapted to darkness, nocturnal animals are acutely sensitive to light; they have evolved mechanisms to protect the eye during daylight, the most important of these being contraction of the pupil. The pupil is the hole in the center of the iris, which appears black because light entering it is absorbed by the tissues of the eye. Pupil size determines the amount of light entering the eye, and is controlled by involuntary contractions and dilations of the sphincter and dilator muscles of the iris—the pigmented light-resistant barrier that regulates the amount of light entering the eye by opening and closing the pupil. Red-eye in flash photography is the result of the iris being unable to close quickly enough, thereby illuminating the blood-rich retina.

Pupil shape is related to a number of factors, including the optical characteristics of the lens, the sensitivity of the retina, and the animal's visual requirements. Pupils are mostly circular, which is the most effective form for maximum dilation at night, and for projecting a clear image regardless of the direction of the incoming light. Coupled with a wide cornea this increases the field of view while the head and eyes remain still. However, the circular pupil is the least efficient design for closing quickly or completely, due to the iris muscles bunching as they contract, and generally is less efficient at blocking light by day, while still allowing a degree of vision. Other pupils have evolved that are more successful, such as the narrow slit pupil, which allows light to be cast over a larger part of the retina than would a tiny circle, providing better daylight vision while still protecting the lens. Slit pupils with both sides coming together are also considered more efficient at closing quickly and almost completely. There are several forms of slit pupil— horizontal, vertical, and occasionally diagonal—the most effective being the vertical one in conjunction with eyelids that can be partially closed at right angles to the slit when the animal squints, thus reducing the length of the opening. There are also some more unusual shapes of contracted pupils such as the triangular ones of the fire-bellied toads, and the stenopaic or lobed pupils of some nocturnal geckos, in which the pupil closes vertically but leaves a few small holes in a line, each one focusing an image on the retina and creating sharp overlapping images; this allows these geckos to hunt in daylight also.

Color perception in mammals is achieved through color receptors, which have different spectral sensitivities. Cones have light-sensitive protein pigments near their surface called opsins which are sensitive to different wavelengths. Most mammals have just two kinds of opsins—the blues and the greens—and therefore have limited or two-color vision. Old World primates have good color vision, however, because they have three types of opsins that absorb light of different wavelengths which are then processed by the brain to produce color images. They are receptive to the red, green, and blue light frequencies and are therefore said to have trichromatic vision, which is believed to have originated in the prosimians or primitive primates. In the New World primates only the females can see color due to the presence of two X chromosomes, which have the red and green vision genes, whereas males only have one X chromosome and therefore have only one copy of each color gene and are color blind.

It is now known that many more animals than previously thought can see color in varying degrees, because they have some receptor cones, although only of the blue and green types, so their color perception is limited. Most diurnal lizards and even the nocturnal geckos have multicolored oil droplets, called retinal filters, in the photoreceptors of their retinas, which are known to be beneficial for color vision, as they increase the range of discernable color, so it is assumed they have some degree of color vision during daylight. Nocturnal animals, including the geckos, cannot see color at night, no matter how many cones they have (and most do not have many, if any), for it is the rods that are involved with night vision, so the world is seen in monochrome. Many nocturnal animals also possess a tapetum, a mirror-like membrane that replaces the black light-absorbing layer behind the retina of the diurnal mammals. This considerably increases the night animal's vision as it reflects back the light that has already passed through it, giving the photoreceptor rods another opportunity to absorb the image. This reflection causes the "eye-shine" seen at night when an animal is "caught" in a car's headlights. Reflected light from the tapetum in nocturnal animals indicates the color of their retinal pigment. Cats reflect gold, bears red or orange, and raccoons yellow.

Excluding the snakes and some geckos, which have an immovable, transparent window covering the eye, many terrestrial vertebrates have three eyelids. The third one is an almost transparent skin called the nictitating membrane, which moves at right angles to the top and bottom lids from the inner corner of the eye. Most amphibians and the other reptiles—lizards, turtles, and crocodiles—possess nictitating membranes, which protect the eye and still allow vision where possible underwater or underground. Most carnivores have third eyelids, while in the prey animals the third eyelid is an important aspect of their protection from predators as it helps to conceal the iris, which is often bright yellow and quite prominent. The nocturnal owls also have third eyelids, with which they can protect their eyes, individually if necessary, against bright sunlight. The human nictitating membrane remains as a tiny piece of pinkish skin in the corner of the eye.

In most birds the bottom eyelid moves upward to close the eye, whereas in mammals the top lid moves down. In most vertebrates a bony socket surrounds and protects the eye, but in some amphibians the eye socket is bottomless, allowing the eyes to bulge below the roof of the mouth and to be depressed in the act of swallowing to help force prey down the gullet.

Hearing

Many nocturnal animals rely upon their sense of hearing as much or more than on their vision, and the ear reaches its highest degree of development in the mammals, with hearing in most species improved by the possession of an outer ear or pinna. The nocturnal species that rely most heavily upon their hearing, such as the bush babies, fennec fox, and the insectivorous bats, have greatly enlarged pinnae, which collect and funnel sounds into the ear canal or auditory meatus and help the animal localize the source of the sound. The depth of the canal and its twisting route protect the eardrum or tympanic membrane and this part of the ear, which is outside the auditory receptors of the inner ear, is known as the external ear. Mammals have external ears, and so do birds, even though they lack outer ears. In birds the sound waves are directed into the external ear by specialized auricular feathers that surround it and funnel sounds toward the auditory meatus.

Fennec Fox *A fennec fox rests in the shade until nightfall. A large-eared nocturnal carnivore of the deserts of North Africa and the Middle East, it relies mainly upon its highly developed hearing to locate the rodents, insects, and scorpions that form the bulk of its diet.*
Photo: Courtesy Ralf Schmode

Despite the size of the outer ear and the presence of an external ear, the most important parts of the hearing sense are the middle and inner ears, because these are the reception and transmission centers of the stimuli, which occurs in all land vertebrates except the salamanders, who only have an inner ear. This part of the hearing system begins at the eardrum, which, in the frogs and toads and reptiles (other than the snakes), lies on the surface of the skull, often in a shallow depression.

Beyond the eardrum the middle ear contains three tiny bones (the malleus, incus, and stapes, with the malleus attached to the inside of the eardrum), which transmit vibrations from the eardrum to the inner ear. The middle ear is connected to the naso-pharynx by the eustachian tube. The snail-shaped cochlea in the inner ear is the actual sensory organ of hearing, containing sound receptors, called hair cells because of the tuft of fine bristles at their ends. They send out electric impulses via the number VIII nerve to the cerebral cortex where they are interpreted. The inner ear has a similar basic plan in all the vertebrates, but it is more complicated in the higher forms, which therefore have a better sense of hearing. Snakes lack external and middle ears so they cannot hear in the human or anthropomorphic sense, but they are far more sensitive to vibrations, the intensity of which informs them whether they are made by a possible predator or potential prey. A small bone (the columella) connects the jawbone or quadrate bone to the inner ear canals, and snakes sense vibrations while their head is on the ground; but airborne sounds are also transmitted from the skin to the jawbone. The salamanders, which lack both an eardrum and middle ear, hear low-frequency sounds through their skulls. Lizards are also sensitive to ground vibrations through their quadrate bone, which articulates the lower jaw with the skull.

Sounds are vibrations in the air, the sound "waves" travelling from their source to the ear where they are captured, and then to the brain where they are interpreted. The vibrations vary in their frequency (the more rapid the vibrations the higher the frequency) and animals also vary in their ability to receive them. They are measured according to their frequency as vibrations or cycles per second (cps) or as units of Hertz (Hz), 1cps equalling 1Hz, and 1,000Hz equalling 1kHz. Currently, it is thought that no animal can hear all sounds, and each has a specific range of hearing that can vary from a few vibrations per second to many thousand.

The human range of hearing commences at 20Hz, below which hearing is replaced by the sense of touch, and extends to 20,000Hz (or 20kHz), beyond which man is deaf. Within their range of hearing some animals may be more sensitive to a particular zone or area. Cats, for example, are most sensitive to sounds in the higher reaches of their range, which extends to 65,000Hz, and can therefore hear a mouse's high-pitched squeaks. Among the mammals the bats have the most sensitive ears, able to pick up frequencies almost to 100,000Hz, and they themselves produce sounds inaudible to the human ear without artificial aids. Rats and mice are almost as sensitive as bats; dogs can hear sounds up to 40,000Hz, double the human range, and can therefore hear the shepherd's "silent" whistle. Sounds above 20,000Hz are termed ultrasonic, and below 20Hz they are infrasonic and are felt rather than heard. Several birds have acute hearing, especially those that rely on echolocation to navigate in pitch-black caves, as they must rapidly receive and process their own calls as they rebound from objects, in time to avoid collisions. Owls rely more on their hearing than sight in the woods at night, with greater reception at some frequencies than any other birds, and are able to locate rodents by their tiny rustling sounds.

Gregarious nocturnal animals rely more upon sound to communicate with their group members or to stake out their territorial boundaries, than do their daylight counterparts, where vision may be more important. Distinctive voices are characteristic of many nocturnal species and it naturally follows that their sense of hearing must be well developed. The hooting and screeching of the nocturnal owls, the roaring of crocodiles and lions, the night calls of migrating waterfowl, and the loud bark of the tokay gecko are all examples of nocturnal communication. There is added need for powerful lungs in the tropical forests, and the booming calls of the monkey-like douroucouli are out of all proportion to its small size. Solitary animals are, as a general rule, less vocal, even though they may possess the faculty, as the need to communicate with their kind is less regular. But the sounds made by nocturnal animals as a means of expression are not confined to those of vocal origin. Other examples of nocturnal communication include the salamander's nasal "snorts," the clapping of owl's beaks, the teeth-gnashing of the peccaries, and of course the rattlesnake's tail rattling. They can also be produced by physical action such as the foot-stomping of the crested porcupine, the deer's hoof-drumming and the bull elk thrashing bushes with his antlers.

Smell and Taste

The sense of smell is very important to many nocturnal animals, for finding food, locating a potential source of danger or a mate, and to smell pheromones—excreted by animals in their feces and urine and by their scent glands—which tell the recipient many personal facts about the sender, including age, sex, and social status. Smell, and the closely allied sense of taste, are chemical senses, so called because scents are chemical compounds. Smell is sometimes described as the first stage of chemical discrimination, providing information in advance and from a distance on factors that may affect an animal or its environment. Particles carried in the air are detected by the scent or olfactory organs and then related to predator or prey, or possible mates or intruders of the same species. Coupled with the stimulation of the olfactory organs there is therefore a physical response, either toward or from the source of the smell. The direction and distance of the source can then be ascertained by the resulting increase or decrease in its strength.

The sense of smell involves the collection of airborne scent particles by receptor cells and their relay to the olfactory lobe and the neocortex of the brain. The receptor cells are located in two clefts situated in depressions above the air passages in each nostril. These contain a network of small bones called turbinates and an epithelium containing millions of receptor cells, which pick up scents as an animal breathes or sniffs. The degree of smell depends upon the size of the nasal epithelium, which in turn varies according to an animal's lifestyle. The larger the epithelium, the more receptors and the greater the sense of smell, and in animals that rely on smell, such as the wolf, tenrec, and paca, the epithelium is very extensive compared to species that rely more on sight. The wolf, for example, has 250 million receptor cells compared to man's 5 million. The olfactory lobe in the front of the brain is responsible for the collection and analysis of the chemical molecules, with the information then being transmitted to the neocortex—the outermost convoluted layer of the forebrain.

In the mammals the olfactory nerve endings are situated in the upper part of the nose, the lower part being concerned mainly with warming and filtering the air they breathe. Consequently, mammals hold their heads high and sniff to locate a source of smell, rather than breathing normally, as this draws more air and therefore more scent particles to the back of the nose. The sense of smell is most highly developed in nocturnal mammals, especially in the carnivores and insectivores.

Birds' olfactory lobes are very small, suggesting little or no sense of smell, and although most nocturnal birds do rely on vision and hearing, smell plays an important part in the lives of some species, especially for food acquisition and nesting. The kiwis have the most highly developed sense of smell in birds, with nostrils located at the tip of the beak, unlike all other birds where they are at its base, and by probing with their beak they are able to locate worms by their smell. The night-active members of the *Procellariformes*—the shearwaters, petrels, and storm petrels—are attracted to fish offal and oil on the sea's surface from several miles away, and are believed to also locate their nest burrows by smell.

Reptiles and amphibians have well-developed senses of smell, with the highly effective vomeronasal system in addition to the standard sense. In the snakes and many lizards this extra sense is associated with the constantly flicking tongue, which picks up scent particles and transfers them to sensory organs in the mouth. Smell is the most important sense for some salamanders that are mute and deaf to airborne sounds.

The sense of taste is considered the second stage of chemical discrimination, and is the ability to detect and respond to dissolved molecules and ions, as opposed to smell, which detects airborne molecules. Whereas smell is the olfactory response to stimuli, taste is the gustatory response and is primarily related to determining edibility. However, the two senses are closely connected, as taste is partly smell and much of the taste sensation comes from odor molecules in the food, since, as we know, certain smells can be tasted. Chemoreception, which is the combined senses of taste and smell, is considered the most primitive of the senses, meaning the first to evolve long ago. Taste is detected by the taste buds, which are groups of sensory cells bearing short "taste" hairs that are stimulated by higher concentrations of chemicals than are the olfactory organs, and result in secretions of the salivary and mucus glands, the act of swallowing and responses in the alimentary canal. Taste receptor cells have a short life (about one week) and are continually being replaced; unlike smell receptors, which are directly connected to the brain, taste buds rely on indirect nerve links to carry their messages to the brain.

Taste is less highly developed in mammals than the sense of smell, and for humans smell refines taste, proven by the difficulty in tasting foods when the nose is "blocked" by a cold. Traditionally, it was believed that the human tongue could detect five basic tastes—sweet, sour, bitter, salty, and the recently discovered umami or glutamate, which is triggered by monosodium glutamate. However, it is now believed that the taste receptor cells may not be specific for a particular characteristic, but may respond to other tastes as well. It is unclear if all mammals have the same tasting ability as man, for in only a few instances are their actual complement of taste buds even known. The dog, for example, has 1,700 taste buds compared to man's 9,000 and the rabbit's 17,000, and is therefore less sensitive to certain tastes. Even less is known about the sense of taste in birds, reptiles, and amphibians, except that in some of the latter taste and smell are important senses, as they have chemoreceptors in the mouth (in the oral cavity and the pharynx) and all over their bodies that are sensitive to chemicals in the air and water.

Birds have fewer taste buds than mammals (with most species having no more than fifty), and so are unlikely to discern a great range of flavors. Many tetrapods (four-legged land vertebrates) are also able to "taste" the air, and the food in their mouths, with their Jacobsen's or vomeronasal organs, which are olfactory receptors containing sensory neurons that detect chemical compounds. They are connected to the olfactory nerves, which convey the information to the brain; and they are present in mammals, and also amphibians and reptiles, where the forked tongues of the snakes and some lizards carry scent particles to them. These sensory organs are absent in birds, primates, crocodiles, most of the turtles, and some bats, and they are most well developed in the monotremes, marsupials, insectivores, and carnivores. In the amphibians they are in a depression adjoining the nasal cavity, and in reptiles in a pit to which the tongue and oral membranes bring chemicals. In mammals they are present in an area of the olfactory membrane inside the nose, which is connected to the mouth by a duct. Current research appears to disprove the old belief that humans no longer possess this second sense of smell, and it seems quite likely that we may still have a functioning vomeronasal organ.

Touch

The sense of touch is very important to many nocturnal species. It is well developed in burrowing animals and in the amphibians that hide by day in or under anything offering shade from the sun. They are said to be positively thigmotactic due to their need to be in close contact with their surroundings. Insectivores such as the hedgehog, moonrat, and solenodon locate their food with their highly tactile snouts, which are strong enough to probe the soil for invertebrates but have a delicate sense of touch. The raccoons have very sensitive fingers with which they search for crayfish and other invertebrates beneath stones at night in shallow water; and the aquatic amphibians, such as the pipa toad and clawed toad, have delicate, elongated fingers that aid them in locating and seizing their food. Whiskers are important tactile organs for several animals, especially the civets and cats, because they extend their touch capability beyond the epidermis. All vertebrates have general somatic receptors in the skin, for touch, pressure, pain, and temperature changes, which are stimulated by contact and undergo electrochemical changes when they are deformed by pressure, sending a message to the brain. In the primates Merkel cells are found where sensory perception is acute, such as in the fingertips; and Meissner cells react to steady pressure on the epidermis.

Birds have a highly developed sense of touch, most of their body surface being covered with sensitive nerve endings and touch receptors. Waders and woodpeckers have tactile sensors, called Herbst corpuscles, in their bills for probing in water or wood, and the insides of birds' mouths and their tongues have similar cells and are especially sensitive. Most insectivorous species have rictal bristles, which are small, stiff feathers around their beaks that act as a net to direct insects into the wide gape. In the nocturnal species all members of the *Caprimulgiformes* except the nighthawks have rictal bristles. Birds also have vibration sensors in their legs with which they detect the approach of a predator on the ground or along the branch on which they are perching.

Notes

1. Elephants in Thailand bolted to higher ground just prior to the 2005 tsunami hitting the shore. It killed thousands of people in that particular spot. All the elephants survived.
2. Obvious exceptions are the owl's acute hearing and the kiwi's sense of smell.

2 Life in the Shadows

When the first fish-like animals crawled out of the water to begin life on land about 350 million years ago in the Late Devonian Period, they set into motion a dramatic chain of events, because this daring change in behavior demanded major adaptations. Lungs had to be developed to replace their oxygen-extracting gills, fins transformed into limbs; salivary glands evolved and eyes that were adapted for underwater viewing became suited for life on land, complete with eyelids. More reliance had to be placed on other senses as they could no longer depend on the receptors along their sides, called lateral line organs, which sensed changes in water pressure caused by predator or prey.

However, these first land animals, which we know as amphibians, did not totally give up their gills (which extract dissolved oxygen from the water), and during the course of their lives most of the 6,000 modern species pass through a gilled stage;[1] with few exceptions they lose them upon becoming adult. The exceptions are aquatic species that have gills throughout their lives, a condition known as neoteny, which may be due to genetics or environmental conditions. But the amphibians' simple lungs did not solve the problem of acquiring sufficient oxygen, and they still rely on skin respiration, in which blood vessels near the skin's surface absorb oxygen from the water or air, for at least half their needs. Lacking gills and lungs, the lungless salamanders are dependent on skin respiration, plus breathing through their mouth linings, and their thin skins have a rich supply of capilliaries which almost reach the surface. Skin breathing requires a moist skin.

Amphibians are of two basic kinds, those with tails, which are called salamanders, and the tailless frogs and toads, which are collectively called Anurans, after their order *Anura*. However, while the name amphibian refers to the ability to live on land or in water, and most do begin life in the water and end life on land, it is not true that all amphibians are equally at home on the land and in the water.

Leopard Frog *Having replaced its gills with poorly developed lungs when it completed the aquatic tadpole stage of its life cycle, the semi-aquatic leopard frog still breathes through its skin when underwater.*
Photo: Bruce MacQueen, Dreamstime.com

Several, such as the eel-like amphiuma and the hellbender, and all fifteen species of pipid toads, including the Surinam toad (*Pipa pipa*) and the African clawed toad (*Xenopus laevis*), are purely aquatic and cannot survive long out of water, although the latter has learned to estivate in a moist cocoon in the mud when its pool dries up. Despite their varying environments and habits, however, they all have one thing in common—the need to remain moist throughout their lives, for all stages of their lifecycle.

Maintaining a moist skin is the most important consideration for amphibians. They cannot tolerate environmental conditions that would dry their skins, and those that live in seasonally arid regions survive by storing water in their bladders, burrowing into moist soil, or producing a protective covering to prevent dessication. A moist skin is necessary for the absorbtion of oxygen and the elimination of carbon dioxide. Oxygen passes through the permeable amphibian skin directly into the blood stream, which transports it to the tissues, and carbon dioxide is returned to the surface of the skin. However, as they can breathe through their skins, their own body fluids are easily lost in the other direction if conditions are unsuitable, and although some can tolerate brackish water, none can live in seawater, where their own body fluids would be drawn out by osmotic pressure. Moisture is especially important to the amphibian egg, to the extent that many frogs make bubble nests to protect their eggs and some species even urinate on them to prevent them from drying out. Keeping moist at all times means avoiding the dessicating effects of the sun, and the most effective way to achieve that is a nocturnal lifestyle. Cool, damp nights are more suitable for amphibians than hot, dry days.

To complement their nocturnal habits and avoid exposure to the sun and its drying rays, amphibians must seek a hiding place at dawn each day, in or under anything that offers shelter. Damp soil and leaf mould, rotting tree stumps, and deep rock crevices are all favored daytime hiding places, especially for the terrestrial salamanders. The large, aquatic eel-like salamanders such as the amphiumas and sirens hide in the mud at the bottom of their ponds, lakes, and rivers during daylight. Frogs also seek wet places, in ponds, marshes, and water-filled ditches, and the delicate-skinned tree frogs may spend the daylight hours attached to the underside of large leaves, where the humid microclimate protects them from dehydration. Many toads spend most of their life underground, appearing only at night after a heavy rainfall. Those that burrow into the soil themselves have evolved specialized adaptations for digging, such as the spadefoot toads (*Pelobatidae*), which have a spade-like knob on the hind feet. Others, like the gopher frog (*Rana capito*) of the coastal plain of eastern United States, hide all day in ground squirrel burrows.

Amphibians can only live in arid regions if seasonal rains allow them to feed and breed periodically or where there is permanent water in oases or waterholes. They cannot survive among the sand and rock of arid regions where it rarely or never rains, such as Chile's Atacama Desert or Saudi Arabia's Rub Al Khali. To maintain the secretions of their mucus glands and keep their skins moist they must absorb moisture, but their adaptations enable them to survive in very arid regions if it rains occasionally. Therefore, in addition to escaping from the sun on a daily basis, there are times when amphibians must hide for longer periods, not just from the direct rays of the sun but from its effects upon their habitat. Migration is out of the question as amphibians are not great travellers, so they must hide for weeks or months at a time. The answer is to estivate—the summer equivalent of hibernation—which generally means digging in, but they must prevent themselves from drying out while they are buried, and this requirement has produced some unusual modifications.

Finding moist soil or a permanently damp place under a large rock is the answer, but this may be impossible in regions with long arid seasons. Where it is likely that their immediate surroundings will dry out completely, amphibians must burrow and provide their own moisture and energy, and protect themselves against its loss. Their water conservation strategy includes accumulating urea in the blood, which raises its osmotic concentration and prevents water passing out through the skin. Species that sleep for several weeks or months survive on their stored body fat and water, but some can manage without food for much longer periods, such as the laboratory toads that lived for two years without food when they were kept moist and cool. The most unusual adaptation to prevent dessication is practiced by Australia's water-holding frog (*Cyclorana australis*), which stores water in its bladder and in pockets in its skin. The aborigines use these bloated frogs as an emergency water source, digging them up, holding the frog's rear end to their mouth and squeezing the water out.

In regions experiencing cold winters amphibians must also protect themelves from frost, and once again the answer is dig in, this time to hibernate—the winter equivalent of estivation—until the arrival of spring. Temperate-climate amphibians hibernate underground in burrows, in the mud at the bottom of ponds, or under leaf mould which is then covered by a blanket of snow. Some can withstand being frozen for short periods, but generally the hibernating amphibians must get below the frost line or be insulated by several feet of snow. Vertebrate animals can only survive freezing if the ice crystals in their bodies are restricted to the extracellular body fluids such as the blood plasma and urine. Water within their cells must not freeze as this damages the cell structure beyond repair. Cold-blooded animals, which can survive freezing for short periods, have evolved specific adaptations that protect their cells. Several northern temperate frogs can withstand temperatures down to 17.6°F (−8°C); and the wood frog even lower, possibly down to 10.4°F (−12°C), at which time about 65 percent of its body fluids are frozen. While they cannot survive for long at such low temperatures, they have lived for two weeks at 27.5°F (−2.5°C). When they are frozen all muscular activity ceases; they have neither heartbeat nor respiration and the frozen blood cannot circulate and carry oxygen and nutrients to the cells. However, the ice that has formed in their bodies is outside the cells only (extracellular), as the cell contents—the cytoplasm and the nucleus—are protected by glycerol or glucose, which function as cryoprotectants or "antifreeze" and depress the freezing point.

■ THE SENSES

Vision

Many amphibians have good vision, and especially good night vision, and some frogs are excellent judges of distance even in low light. The arboreal "flying frogs" of the genus *Rhacophorus* obviously have exceptionally good long-distance eyesight to glide between trees 50 feet (15.2 m) apart, and can intercept passing insects in flight with their long, viscid tongues. Their horizontal pupils expand greatly in poor light, enabling them to live in the forest canopy where the dense foliage blots out most of the moonlight. In contrast the terrestrial toads have poor long-distance vision, and experiments have shown that some cannot see movement beyond 10 feet (3 m). The frogs and toads, especially the terrestrial and arboreal species, rely on sight when hunting their prey, its movement being the stimulus that triggers the seizing response.

Frogs' eyes project from the skull and give them a clear view of approaching food or potential danger, as they can see in all directions. Most do not have binocular vision at very close range, and when prey suddenly appears under their noses they must back away to bring it into focus. They are also color blind and do not recognize red any more than a fighting bull. Despite the good eyesight of many species, however, the amphibian eye is not a highly developed organ. The lens is fixed[2] and the upperpart of the frog's eyeball is protected by a thick layer of skin which is actually the immovable upper eyelid, and in some species the lower lid has become a nictitating membrane that can be drawn up to protect the eye while still allowing vision. To close its eyes a frog must pull them down into their sockets and then cover what still shows with the nictitating membrane.

Gladiator Tree Frog *Large eyes projecting from its skull and enormously enlarged pupils in darkness provide this neotropical tree frog with exceptional night vision as it searches for insects while clinging to smooth surfaces with its toe pads.*
Photo: Lynsey Allen, Shutterstock.com

The amphibian eye shows a great variety of iris coloration, which does not seem to have any special purpose because it may resemble the animal's body color, or be in total contrast like the red-eyed tree frog, whose body is bright green. Frogs have large pupils but salamanders generally have even larger ones, although with smaller and less colorful irides, and expansion of the pupil is so great on dark nights that very little iris is visible. When the iris contracts during daylight to reduce the size of the pupil and protect the eye, the aperture in most species is a horizontal slit, while the spade-foot toads (*Pelobates*) are distinguished by their vertically elliptical pupils, and some of the fire-bellied toads (*Bombina*) have an unusual, triangular-shaped pupil.

Amhibians not only have good vision but also have light-sensitive skins, proven in laboratory experiments in which blinded frogs in a darkened box moved to face the direction from which light was allowed to enter. In addition to the sensitivity of their skins the amphibians, and many other animals too, have the remains of a third eye. The long-extinct ancestors of the frogs and toads actually had three eyes, the extra one being the pineal or parietal eye, positioned on top of the head. What remains of this eye is visible in the larval stages of some amphibians but is no longer functional, as it is in some lizards.[3] This third eye was a photo-sensitive organ concerned with light reception, especially changes in light, and did not discern images, but it degenerated as the amphibians became nocturnal and light reception was superfluous. The pineal gland remains deep in the brain, however, and melatonin—the hormone it produces—is involved in the thermoregulation of the amphibians (and many other animals including humans) and plays an influencing role in the daily rhythm of activity.

Frogs' eyes have another valuable purpose in addition to vision—they help them swallow their food. Their eyeballs project below the roof of the mouth, and they swallow by closing their eyes and depressing the eyeballs, which helps to force their food, with the help of their tongues, down the gullet. Ingestion in frogs involves a lot of gulping. Except during the early stages of their lives, when as tadpoles they are most likely to eat plant matter, amphibians are totally carnivorous and rely upon animal protein for their energy. Movement stimulates their optical senses, and their response depends upon the size of the object, large ones repelling them, smaller ones triggering the seizing action. They are sensitive to the slightest movement of their prey, and a toad's attention can be drawn to the movement of the rib cage of an otherwise perfectly still mouse.

Hearing

Amphibians lack an outer ear or pinna which channels sounds into the ear canal, and is the organ usually most associated with hearing. They also lack an outer ear canal or auditory meatus that in mammals leads to the eardrum or tympanum, which separates the outer ear from the middle ear. Hearing therefore begins in the frogs and toads at the eardrum, which is the exposed flat and shiny round patch just behind the eye. This is not true of salamanders, however, which lack an eardrum and a middle ear, and therefore probably only "hear" low-frequency sounds transmitted through the skull.

The tympanum covers a funnel-shaped cavity that connects to the otic capsule, which encloses the inner ear. This capsule also has a passageway to the pharynx, called the eustachian tube. Sound waves therefore reach the frog's tympanum from outside, directly onto the tympanum's outer surface, and via the mouth and eustachian tube onto its inner surface. What most amphibians do have in common is a well-developed internal ear, comprising a middle ear and an inner ear, and in the frogs and toads hearing is quite acute. Their middle ear is the cavity behind the eardrum, but it has only one bone (unlike the mammals, which have three—the hammer, anvil, and stirrup) to transfer sounds to the inner ear. The inner ear contains the cochlea, a spiral structure that converts sounds from mechanical vibrations to electrical signals, a process known as transduction, and these signals are carried to the brain by the auditory nerve.

The sense of hearing in amphibians is associated with the possession of a voice, and both are well developed in the frogs and toads, which have a variety of calls, often quite loud. One mute exception is the tailed frog (*Ascaphus truei*) of the Pacific Northwest, which lives in cold streams where it clings to rocks with its sucker-like mouth. Hearing is tuned to receive the calls of their conspecifics, so they can locate potential mates in a pond where several species may be calling at once. Each has a distinctive call that is recognized by others of the same species, and there is considerable specific variation in frequency, duration, pitch, and volume. The dusky gopher frog (*Rana sevosa*) has a deep snoring call, the leopard frog (*Rana pipiens*) makes a guttural trill, the southern cricket frog (*Acris gryllus*) makes a rapid rattling sound like a metal clicker, and the bullfrog (*Rana catesbiana*) utters a series of sonorous bass notes.

Practically all species attract their mates by vocalizing, but the calls heard at night are those of the males, as the females of most frogs and toads are mute. Experiments have shown they are sensitive to sounds ranging from 50 Hz to 10,000 Hz, and they are also sensitive to ground vibrations, a single footfall being sufficient to silence a pond full of croakers. Frogs have a voicebox or larynx, and in many species the males have an expandable throat sac which acts as a resonator and helps them make their loud and varying sounds. Some can produce sounds with their mouths closed, so that they can actually vocalize underwater and even underground. The aquatic clawed toad (*Xenopus laevis*) and Surinam toad (*Pipa pipa*) lack both vocal sacs and tongues, but both sexes make clicking noises with their mouths while underwater.

The volume of a frog's call and its hearing sensitivity are related to the urgency with which mates need to be attracted to take advantage of special circumstances, usually rainfall and the chance to breed. Spadefoot toads (*Pelobates*), which have loud, far-carrying calls, live in relatively dry country and need to breed rapidly after a sudden downpour. In addition, their tadpoles mature very quickly and can metamorphose into tiny toads two weeks after hatching. At the other end of the volume scale the red-legged frog (*Rana aurora*) of western North America makes a barely discernable grunt, because it is a permanent water-dweller. To increase their chances of finding mates some frogs even have a distinct breeding call. Florida's barking frog (*Eleutherodacytlus augusti*), for example, reduces its many-syllabled tree-top call to a single bark when seeking a mate down at pond level.

Unlike the often loud calls of the noisy frogs and toads, the salamanders have no voicebox and with relatively few exceptions are mute. The Pacific giant salamander (*Dicamptodon tenebrosus*) of western North America manages a low rattling with its lungs and mouth, and the arboreal salamander (*Aneides lugubris*), one of the lungless or woodland species, makes a weak squeak by contracting its throat and forcing air through its nostrils. It is not surprising that the mute salamanders cannot hear in the accepted sense, for they lack an eardrum and are therefore deaf to airborne sounds; but they are very sensitive to ground vibrations through their bodies, and to airborne sounds below 250 Hz.

Low-frequency vibrations are also picked up by amphibians in the water through their sensitive lateral line organs, retained from their ancestors, although they are less sensitive than those of the fish. They are groups of sensory cells that respond to low-frequency vibrations and relay the information to the central nervous system. They contain receptors known as neuromasts which have hair-like projections that sense low-frequency vibrations, informing the animal of even the slightest movement in the water nearby. Lateral line organs are present in the aquatic salamanders and the pipid toads throughout their lives; they are also present in tadpoles of the terrestrial anurans, but are lost during metamorphosis as they are not needed on land.

Smell and Taste

Amphibians have a good sense of smell, which involves using the cells inside their simple nostrils and the associated olfactory organs, where smells are registered and analyzed. This arrangement is also complemented by Jacobsen's or vomeronasal organs—the extrasensory cells in the roof of the amphibian's mouth—which intercept scent particles and transmit information to the brain for analysis.

A good sense of smell is most important to those species with poor vision, for the burrowing amphibians, for example, in their search for invertebrates underground, and for the nocturnal aquatic species in dark water. Even in some terrestrial amphibians smell is well developed, as in the natterjack toad (*Bufo calamita*), which is attracted to garbage dumps by its keen sense of smell, although only to eat the cockroaches living there; and the marine toads in Trinidad, which raided our dog's bowl at night for dog food. But these are exceptions, for movement stimulates them and smell does not generally induce a feeding response in amphibians other than the salamanders. Most amphibians have no interest in inanimate objects and would starve to death alongside a dead mouse or a pile of dried flies, although some captive specimens have been induced to accept inanimate foods, and some, such as the aggressive African bullfrog (*Pyxicephalus adspersus*) and horned frog (*Ceratophrys ornata*), need little inducement to eat dead mice.

The sense of smell is undoubtedly more important to the salamanders, as their larvae proved when they had their nostrils blocked experimentally in the laboratory, and were then completely insensitive to food. Mute, and deaf to airborne sounds, salamanders also rely mainly upon their sense of smell to attract or locate mates, and even the lungless species have nostrils and well-developed olfactory organs, plus glands on their chins that secrete pheromone-rich mucus to attract females at breeding time. Salamanders taste with their tongues, and in the frogs and toads the palatability of a beetle or woodlouse is determined by the taste buds which line their mouths, and they quickly reject distasteful items or inanimate objects such as grass or soil caught up with their food.

Touch

In the completely aquatic Anurans the sense of touch is well developed. Clawed toads and Surinam toads act like racoons, searching for food beneath stones and in mud and dense weed with their delicate fingers, which have tiny sensors at their tips. The tongue-less Surinam toad also holds it hands close to its face, spreading out its long fingers on either side of its mouth, and attracts prey by vibrating a small flap of skin hanging from its snout. When a fish comes within reach of the nerve endings in the fingers a message is flashed to the brain and the toad seizes it or lunges forward with open mouth to siphon it in. Their fingers are so important that if one is lost it can be regenerated. The amphibians are also well aware of occurrences in their environment, as their skin contains a network of uncoated nerve fibers that serve as receptors of heat, cold, pain, and pressure.

■ SOME OF THE SPECIES

Salamanders

The order *Urodela*, whose members are collectively referred to as salamanders, contains two disinct groups of animals—the newts and the salamanders—which are also known separately by those names, an unfortunate situation further confused by the fact that newts are members of the family *Salamandridae*, which also contains the fire salamanders. The salamanders are mostly northern, temperate-climate animals, of both the Old and New Worlds, although they are especially well represented in North America, and there are none in Australia. They have long tails and cartilaginous pectoral girdles, and although they all have limbs, in some species these are tiny and useless and in others are present as forelimbs only. Their vertebrae range in number from twelve to sixty-two. The larvae of most species have external gills, and some retain these throughout their lives, but others have neither lungs nor gills and breathe only through their skins or mouths. Some salamanders give birth to living young, but most lay eggs, and fertilization is usually internal, unlike the frogs and toads, where the eggs are fertilized after they have been laid in the mating arrangement known as amplexus. Most salamanders have smooth and slimy skins due to the mucus secretions of their glands, and like the toads and some frogs their skins also secrete toxins, some of them being extremely noxious and even fatal to humans if swallowed. The salamanders are either terrestrial, purely aquatic, or they practice a combination of land and water living, although many of the land-dwellers lay their eggs in water and their larvae are aquatic.

Terrestrial Salamanders

Tiger Salamander (*Ambystoma tigrinum*)

One of the mole salamanders, the tiger salamander occurs across most of North America from southern Canada to Mexico, excluding the Great Basin and southwestern deserts and Florida. It is the largest New World terrestrial salamander, reaching a total length of 12 inches (30 cm), with many subspecies and variations of color and pattern, but generally having a dark background color of blackish or greenish-brown with yellowish-brown or olive spots or blotches, and usually a brownish-yellow belly. The tiger salamander is a stocky animal, with a broad snout and tiny eyes. Aquatic only when breeding, adults spend much of their lives underground in rodent holes, migrating in spring to shallow and quiet water to breed. They are totally nocturnal and search for worms and slugs in the soil and in deep leaf litter after nightfall.

Marbled Salamander (*Ambystoma opacum*)

Another mole salamander, but one of the smaller species, the marbled salamander grows to just over 4 inches (10 cm) long. Basically black, it has extensive marbling on its back and sides, in the form of bands and blotches. These markings are usually gray in females and white in males. The marbled salamander is less dependent upon moist areas, and is often found on dry hillsides, but moisture is essential for reproduction. The female lays her eggs in a shallow depression in the fall, and then guards the eggs until they hatch, which occurs soon after the depression fills with water. The larvae are aquatic and feed at night, and when five months old they leave the water, lose their gills, and from then on live on land. This salamander lives in the eastern and southern United States from Massachusetts to Texas.

Fire Salamander (*Salamandra salamandra*)

A native of Europe, North Africa, and southwest Asia, the fire salamander is a large and robust animal that reaches a length of 12 inches (30 cm) including its tail. It has well-developed poison glands concentrated around the head and on the back, which contain the neurotoxic alkaloid samandarin, but is brightly colored to warn potential predators of its toxicity. Although individuals vary considerably in color they generally have bright yellow, orange, or reddish spots or stripes on a glossy black background. They are found in damp forest, rarely far from water, and roam nocturnally following rainfall. A breeding male produces a packet of sperm

Fire Salamander *A terrestrial species when adult, the fire salamander from Eurasia and North Africa is starkly patterned to warn potential predators that its skin glands contain a powerful neurotoxin which can be fatal to small mammals.*
Photo: Carsten Reisinger, Shutterstock.com

which he deposits onto the ground and then grasps the female and lowers her onto it. She gives birth to either well-developed larvae or fully developed metamorphosed young, usually depositing them directly into water.

Semi-Aquatic Salamanders

The salamanders that spend time in the water or on land, in varying degrees, are usually called newts, although the names are often used interchangeably. Most have rougher and drier skins than the ones known as salamanders, and they never have gills when adult. Most newts are aquatic, although many leave the water after the breeding season, and some hibernate on land for several months. The red-spotted newt (*Notophthalmus viridescens*) has a land stage—called the red-eft—where it spends up to three years as a totally terrestrial animal. Newts have toxic skin gland secretions and are rarely eaten by predators. Their costal[4] grooves are much less distinct than those of the salamanders, and males develop smooth skins, flattened tails, and a spiky tail or a total dorsal crest prior to breeding. None of the newts are large animals, rarely exceeding 12 inches (30 cm) in length.

Great Crested Newt (*Trichurus cristatus*)

A large, warty-skinned newt, reaching a length of almost 7 inches (16 cm), it has dark, grayish-brown sides and back, with so many darker spots it appears almost black. There are small white spots on the lower flanks, and the belly is yellow or orange with dark blotches. Only the male has a crest which becomes more pronounced during the breeding season. This is Britain's largest newt, although absent from Ireland, and it also occurs across northern Europe from France to the Urals. The great crested newt spends most of the year in or around ponds and small lakes where it breeds in water, and its larvae are aquatic. It is very carnivorous and cannibalistic, and both the adults and larvae eat tadpoles, worms, and insect larvae, with the adults also preying on small frogs, snails, and even baby newts. Adults pass their inactive daylight hours hiding under logs, large rocks, and also in burrows, and they hibernate from October to early March.

Smooth or Common Newt (*Trichurus vulgaris*)

This newt is a more terrestrial species of mole salamander, a pale-brown or olive-green animal with two dark stripes on its back, and with a black-spotted orange belly, paler in the females. In the breeding season the males have a continuous back crest. It lives in northern Europe and western Asia from France to Russia, where it spends its time on land searching for insects and worms which it catches with its projectable tongue. When in the water it seizes aquatic insect larvae and tadploes with its jaws, which have tiny teeth. The smooth newt spends the daylight hours under rocks and logs and in leaf litter. It hibernates from October to

March in the northern parts of its range, and is quite cold-hardy, emerging when the temperature is still at the freezing point. It is also dependent on water for breeding, laying its eggs on a submerged plant leaf, and initially the aquatic larvae absorb oxygen through their external gills and metamorphose into air-breathing adults at the age of ten weeks.

Rough-skinned Newt (*Taricha granulosa*)

A dark-brown or black newt with a bright-orange belly and a rough and dry warty skin, this species has small eyes and dark lower lids and reaches a maximum length of 5 inches (12.5 cm). Its range is the western coastal lowlands of North America from Alaska to San Francisco Bay, and it is the most aquatic of the Pacific newts. In springtime in the early morning it is commonly seen lying chilled and immobile in the open after a night of rain. These newts arch their backs and raise their tails to show their bright undersides in an attempt to intimidate attackers, but if that fails they have highly toxic skin secretions, which have killed dogs that have eaten them. The rough-skinned newt lives in moist grasslands and woodland in the vicinity of water courses and ponds. In the breeding season—from January to July—the male's skin becomes smooth, and the females lay their eggs in water, attaching them singly to the leaves of aquatic plants.

Aquatic Salamanders

A number of large species of salamanders have been loosely grouped together as "giant salamanders," and these are all totally aquatic and nocturnal. Although they have very interesting habits, the animals themselves are considered rather grotesque. These salamanders are members of four families, the *Proteidae* (mud-puppies and olms), which are permanently larval; the *Amphiumidae* (amphiumas), the *Sirenidae* (sirens), and the *Cryptobranchidae*, two of whose three members live in eastern Asia. These include the very rare Japanese giant salamander, which grows to 5 feet (1.5 m) long, and is the world's largest amphibian.

Hellbender (*Cryptobranchus allegheniensis*)

The sole New World representative of the family *Cryptobranchidae*, the hell-bender lives in streams in the Appalachian and Ozark Mountains of the eastern United States. It has a large flattened head, and slimy, wrinkled folds of skin on the sides of its body, which absorb oxygen from the water. However, appearances are deceiving, since it is quite harmless. Either brown or black with indistinct spots, it has a short and tapering tail which is keeled and used for propulsion. Its record length is 29 inches (74 cm), at which size it may weigh 5 pounds (2.2 kg). The hellbender lives in clear rivers and streams, usually in fast-flowing water, and hides under rocks and logs during the day, appearing at night to search for crayfish, fish, frogs, aquatic insect larvae, and snails. It has external gills only as a larva, and like the Asiatic giant salamanders it has small limbs.

Mudpuppy (*Necturus maculosus*)

A very long-bodied, eel-like salamander, with tiny limbs, the mudpuppy is so named because it, and the related waterdogs, are erroneously believed to bark. It lives in permanent rivers, lakes, and ponds in eastern North America—from southern Canada to the Tennessee River. It reaches a length of about 12 inches (30 cm) and is grayish-brown with scattered blackish spots, and its belly is gray with dark spots. Mudpuppy larvae are usually striped, with a dark line along the spine flanked on both sides by a yellow stripe. Like the axolotl, this species spends its life as a permanent larva, retaining its prominent external gills, which are colored maroon and wave like bird plumes in the water current. It is active at night, searching for crayfish, insect larvae, snails, and fish in the dark waters.

Two-toed Amphiuma (*Amphiuma means*)

The amphiuma is the largest New World amphibian, which commonly reaches a length of 30 inches (76 cm), although the record is 39 inches (1 m). It is a dark-brown or black eel-like amphibian with four tiny and useless limbs and no external gills, and has very slimy skin. It lives in permanent bodies of water in the eastern United States from Virginia to Florida. Highly carnivorous and with powerful jaws, it lies in ambush for its prey in the mud or under pond-bottom debris with its head protruding in the manner of the marine wolf eels, and consequently is often called a conger eel. With its large size and sharp teeth it is capable of giving a serious bite, second to that of a snapping turtle. Although highly aquatic in the still waters of swamps, sloughs, and ponds, it may change location at night by wriggling through swampland.

Frogs and Toads

The frogs and toads are tail-less after they have metamorphosed from tadpoles,[5] and this is the most obvious difference between them and the newts and salamanders. Although scientifically the name "frog" applies only to the family *Ranidae*, and "toad" to species within the family *Bufonidae*, the members of all the other families are commonly called either frogs or toads, and collectively may be called anurans after their order *Anura*. Toads are stouter, often squat animals, mostly with dry, warty skins and short legs more suitable for hopping, and they are nocturnal or crepuscular. Some have large parotoid glands located behind the cranial crest, which produce very toxic secretions. They are more lethargic than frogs, and also more adaptable, having evolved to cope with varying conditions, involving considerable variations in humidity as well as temperature. There are purely aquatic species and terrestrial forms which require water only for breeding.

Frogs have smooth, moist skins and are more dependent upon a wet or moist environment, and in addition to both terrestrial and aquatic forms there are also arboreal frogs. They are more graceful in shape than the squat toads, and have long hind legs for leaping.

With very few exceptions fertilization of Anuran eggs occurs externally; mating involves the sexual embrace known as amplexus, in which the male hangs tightly to the female's back and fertilizes the eggs as they are laid. With the most complex skin of all living creatures, frogs and toads are becoming increasingly interesting to science because of the many chemical compounds in their skins, which have potential value in human medicine.

Typical or True Frogs

There are representatives of these amphibians on all continents except Antarctica. They are characterized by their long legs, pointed toes, and webbed hind feet. Their skins are smooth and moist, with dorso-lateral folds, and they have large and prominent eardrums. During the breeding season, the males grow a dark nuptial pad on the thumb for holding the female during the breeding embrace called amplexus.

Edible Frog (*Rana esculenta*)

A bright dark-green frog, 5 inches (13 cm) long, with black spots on its body and bars on its legs, and with a narrowish yellow dorsal stripe, this species is a native of mainland Europe and has been introduced into England. It is very aquatic and is rarely seen far from water; like other members of the family *Rana* it has paired vocal sacs situated at the sides of the throat. Its hind legs have been a delicacy since the days of ancient Rome, although other *Rana* species are also important as food. Its legs are still eaten, especially in France. The edible frog is actually a hybrid between the pool frog (*Rana lessonae*) and the marsh frog (*Rana ridibunda*). A mating between two edible frogs is usually infertile, but they can produce fertile eggs if mated back to one of the original species.

American Bullfrog (*Rana catesbiana*)

A large frog, varying in color from dark green through shades of brown to dark gray and black, with a whitish or yellow belly, the bullfrog is a native of eastern and central North America. Its most distinguishing feature is a very large and distinctive eardrum or tympanum behind the eye, which is much larger in males. It is highly aquatic, preferring large bodies of water, especially well-vegetated still lakes and bogs. It is also important in the gourmet food trade, being farmed for that purpose in the United States; it is now established in several countries, especially in Europe, as a result of escapes during attempts to farm them commercially. It is the largest North American frog, reaching a length of 8 inches (20 cm), and is able to eat small birds and hatchling turtles.

African Bullfrog (*Pyxicephalus adspersus*)

This is a large frog, in which the males reach a length of 9 inches (23 cm) and weigh up to 2½ pounds (1.1 kg), but the females are only half the size; this is unusual in frogs since the female is usually larger. The African bullfrog lives in pools on the open savannah, scrub country, and bushveldt, in eastern, central, and southern Africa, often in quite arid regions. When the pools dry up it burrows into the soil with its powerful front legs, and spends much of the year underground. It has a very stout body and broad head, and only the hind toes are webbed. Although there is considerable color variation, African bullfrogs are usually olive-green with darker ridges. Males have a creamy-yellow throat, and the female's throat is white. They are very aggressive frogs, capable of swallowing mice, small lizards, and other frogs, with tooth-like projections in the lower jaw that help them to maintain a grip on their struggling prey.

Narrow-mouthed Toads

A family of over 300 species of burrowing, terrestrial, and arboreal toads, found throughout the world, the narrow-mouthed toads have short legs, thick bodies with no waist, and a small pointed head. Unlike typical toads their skins are smooth and there is a fold of skin behind the head. Some species lay their eggs on land, and metamorphosis then occurs in the egg, with the young emerging fully formed. Their pointed noses are an adaptation for foraging for ants and their pupae.

Tomato Frog (*Dyscophus antongilii*)

A very attractive frog in which the female is 4 inches (10 cm) long and the male about half her size, they occur only on the northeastern coast of Madagascar, where they breed in shallow pools and swamps. They are now very rare in the wild due to the loss of habitat and overcollecting for the pet trade, but are now bred frequently by zoos and private herpetologists. They are plump animals with short limbs and pointed heads, and are sexually dimorphic, the females being bright red, and the males a dull yellowish-orange. In both sexes the belly and the tips of the toes are yellowish-white, and they have green eyes. Tomato frogs puff themselves up when confronted by a predator, and if that is ineffective they secrete a thick, white toxic substance from their skin glands (see color insert).

Great Plains Narrow-mouthed Toad (*Gastrophyrne olivacea*)

A small brown or gray smooth-skinned toad, only 1½ inches (4 cm) long when adult, the great plains narrow-mouthed toad has a small, pointed head and a thick waist, and short and stout hindlimbs. Males have small tubercles on the chest and lower jaw. It is a secretive animal, hiding by day in burrows, under rotting logs and rocks near water, in areas of grassland and woodland. Its call, in chorus, resembles a swarm of bees and has been likened to a flock of bleating sheep when heard from a distance. It lives in southern Arizona and Mexico, and from southern Nebraska south into Mexico, on either side of the Continental Divide, dry regions where it is only possible for these toads to breed after the summer rains.

Tree Frogs

The family *Hylidae* contains about 700 species of very smooth-skinned frogs with long legs and big toes that look and act like suction caps. They are usually bright green with markings on their sides or legs, but like the chameleon they are able to change color, and at times they become quite dark. They are mostly small frogs, less than 2½ inches (6 cm) long and are mostly arboreal, but all lay their eggs in water—some tropical species in tree-top bromeliads. The largest species is the gladiator tree frog (*Hyla boans*) of northern South America and Trinidad.

Common Tree Frog (*Hyla arborea*)

A plump little long-limbed and smooth-skinned frog, just 2 inches (5 cm) long, from Europe and Asia Minor, the common tree frog has small, disc-like pads on its fingers and toes. It varies considerably in color, but is usually bright green with a dark stripe from the eye along the flanks to the groin (can also be yellowish-brown to dark-brown with black blotches), and when it emerges from hibernation it is usually brownish-gray. Males have a large yellow vocal sac under the chin. The common tree frog lives in well-vegetated habitat, usually near water; it is a good climber, often going high into trees, and although nocturnal it often sun bathes. It has been introduced to the United Kingdom on several occasions but has not survived—due either to the coolness of the climate or collectors who supply the pet trade.

Marsupial Frog (*Gastrotheca riobambae*)

A native of Amazonian Peru and neighboring Brazil, the marsupial frog is a nocturnal tree frog with mainly terrestrial habits. Females are bright pastel-green above and usually pale golden-brown below, with golden warts on their backs. At 4 inches (10 cm) long they are double the size of the males, which are blotched all over with dark-brown and pale golden-brown. The female marsupial frog has a pouch on her lower back in which she carries and incubates her eggs. As she lays the eggs while in amplexus, the male pushes them into her pouch, aided by a secretion which he beats into a slippery foam with his legs. The tadpoles are eventually released from the pouch when about six weeks old, the mother holding the pouch open for them. She deposits them in water, but must take care, as marsupial frogs are not good swimmers and have drowned in deep water.

Tropical Frogs

This is a large and very diverse family of frogs with at least 800 recognized species. They are mainly neotropical, but some live in the temperate regions of the New World and a few in Australia. Generally, they lack ribs and have teeth in their upper jaws. They include frogs that have adopted fossorial, arboreal, terrestrial, and aquatic habits. Most lay their eggs in water, some make foam nests, and others lay their eggs on the ground and guard them.

Ornate Horned Frog (*Ceratophrys ornata*)

Also known as Bell's horned frog, these large amphibians from Brazil and eastern Argentina grow up to 8 inches (20 cm) long and have squat, round bodies with huge heads and mouths. They are the most carnivorous of all the anurans, and are ambush hunters that lie in wait for mice, lizards, other frogs, and even small snakes; but with their great appetites and a sedentary lifestyle they are high-risk cases for obesity. Horned frogs are terrestrial and have been known to drown in deep water, possibly due to being overweight. They are bright green with large, reddish-black spots and blotches with yellow margins, a large head and very wide mouth, and their "horns" are extensions of the upper edge of their eyelids (see color insert).

Mountain Chicken (*Leptodactylus fallax*)

This large frog, which reaches a length of 8 inches (20 cm) and may weigh 2 pounds (900 g), hails from the islands of Montserrat and Dominica in the West Indies. It has long been a favored item of food and is known locally as the mountain chicken. Overhunting for food and the loss of habitat has exterminated this species from other Caribbean islands, such as Martinique and Guadeloupe, and on Montserrat it is now further endangered by the recent volcanic activity. In Dominica it has survived due to the protection of its mountain forests, where it lives in fast-flowing streams. A foam-nester, it is one of the species in which the seminal fluid released by the male during amplexus is beaten into a foam to cover the eggs, and this hardens on the outside and protects them during their development. Whereas most foam-nesters place their nests over or near water, the mountain chicken creates its nest on land, usually in a burrow, into which the male entices a female by trilling. She stays close to the nest and aggressively defends it, while the male stands watch outside. When the eggs hatch the mother feeds the larva with trophic (unfertilized) eggs.

Typical or True Toads

The true toads have plump bodies and short legs and a dry and warty skin. Their faces are round and they have horizontal pupils, with large parotoid glands behind the eyes that secrete a very toxic substance, sufficiently powerful in some species to kill a dog. These amphibians are distributed worldwide except in Australia and in cold and dry regions, and they have not colonized oceanic islands. Several darken in color in response to temperature changes.

European Green Toad (*Bufo viridis*)

This is certainly the most attractive European toad, a yellowish-white animal mottled with green, with some warts on its back; and the males have nuptial pads on their fingers when in breeding condition. Despite its name, it occurs also in North Africa and in southern and central Asia. Females are a little larger than the males and reach a length of 4 inches (10 cm). The green toad prefers dry and sandy habitat; it often frequents developed areas where it hunts insects attracted by street and house lighting, and only enters water in the spring for breeding. It is not a highly active animal, preferring to sit and let its prey—slugs, worms, and wood lice—approach within reach. It has few, if any, enemies due to the very toxic contents of its parotoid glands.

Cane Toad or Marine Toad (*Bufo marinus*)

These large brown toads, with very warty skins, grow to 9 inches (23 cm) long, may weigh 2½ pounds (1.1 kg), and are able to leap 3 feet (90 cm). Handling them is dangerous because of the creamy poisonous discharge from their large parotoid glands, which is a natural defense mechanism against predators, and has proved lethal when domestic dogs have eaten them. Marine toads are natives of tropical America, but are now established in several countries after being introduced mainly to control insects in the cane fields. They have been particularly harmful to native animals in Queensland, as predators and also as prey—especially to the snakes that eat them. Researchers at the University of Sydney have estimated that 30 percent of all native snake species are at risk of being poisoned if they eat several cane toads, and for seven species, a meal of just a single toad could prove fatal. In Fiji they prey on the young of the very rare endemic banded iguanas. Their common name is misleading as it implies they are also at home in salt water, but they can only tolerate the brackish conditions of estuaries, coastal lagoons, and mangrove swamps.

Colorado River Toad (*Bufo alvarius*)

The largest toad in western North America, reaching a length of 6 inches (15 cm), the Colorado River toad is dark-brown or olive above with a creamy belly. Despite its smooth and shiny skin, it does have several warts, a large, whitish one on each side of the face at the angle of the jaw, and large, warty glands on the legs; the young also have warts but outgrow them. This toad's parotoid glands are large, elongated, and kidney-shaped and touch the prominent cranial crests. The Colorado River toad lives in southern Arizona and northwestern Mexico in arid grasslands, woodland, and in mountain canyons, usually near water (either permanent

Cane or Marine Toad   *This large toad is one of the few amphibians which can survive in the brackish water of estuaries and mangrove swamps. Introduced into Australia, Guam, Samoa, and Hawaii to control sugar cane beetles, it has had a devastating effect on native wildlife.*
Photo: Sander van Sinttruye, Shutterstock.com

waterholes, springs and streams, or temporary pools), but it has been found several miles from the nearest water. It is very nocturnal and is most active when it breeds after rainfall from May to July.

Spadefoot Toads

These small, plump toads differ from the true toads (*Bufo*) in several ways. They lack the large parotoid poison glands; they have only a single sharp-edged "spade" on their hind feet for digging, whereas the true toads have two rounded tubercles; and they have teeth in their upper jaw, whereas the other toads are toothless. They also have vertical, cat-like pupils, which in the true toads are horizontal, and their skins are smooth, not warty. They burrow backward into the soil, where they remain all day and during very hot and very cold periods.

Common or European Spadefoot Toad (*Pelobates fuscus*)

This is a smooth-skinned nocturnal toad from central and eastern Europe, where it is primarily terrestrial and enters ponds and marshes just in spring to breed. It has two distinguishing features: a large, pale "spade" on each hind foot—actually a metatarsal tubercle—with which it burrows into sandy soil; and a lump on top of its head. Reaching a length of 3 inches (8 cm), European spadefoot toads are variable in color, usually yellowish-brown or gray with dark-brown blotches or stripes and small orange spots on their sides. When threatened they inflate their bodies and jump up at their attacker with open mouth. Their tadpoles grow very large, up to 6 inches (15 cm) long, prior to metamorphosing into toadlets.

Malaysian Leaf Frog (*Megophrys nasuta*)

Despite its name this most unusual species is a member of the *Pelobatidae* and is therefore a spadefoot toad. It is a stocky animal with a wide head and short, thin hind legs, and its eyelids and snout are elongated into "horns." Reddish-brown in color, it has smooth skin with two ridges running along its sides, and in shape and color it resembles a large dead leaf, so it is well disguised for terrestrial life in the leaf litter of Southeast Asian and Indonesian forests. Females reach a length of 6 inches (15 cm) and males are considerably smaller. It is totally nocturnal and hides on the forest floor by day, emerging at dusk to feed on invertebrates such as snails and cockroaches. It breeds on the banks of clear, swift-running streams and its aquatic tadpoles have turned-up mouths to feed on surface particles.

Pipid Toads

The *Pipidae* is an entirely aquatic family of toads, found only in Africa south of the Sahara and in northern South America. Like fish they have lateral line organs, which are sensitive to motion in the water and alert the toads to potential prey or predator. They lack tongues, but are able to make clicking sounds through movement of the cartilage in the larynx.

Clawed Toad (*Xenopus levis*)

The clawed toad, which grows to 4 inches (10 cm) long, is usually brown above with a pale belly, but albino specimens are frequently seen. Females are usually a little larger than males. Characteristics of this species are their habit of floating motionless below the surface for long periods, and their energetic amplexus behavior, in which they swim in circles up to the surface where the females lay their eggs. This toad breeds readily and, after being used as a major research animal for many years, is now considered domesticated. Its original value to science was in pregnancy tests, but antibiotics recently discovered in its skin are now being developed. It is now a pest in several countries after laboratory specimens were released into the wild. Clawed toads have very long claws that are used for digging invertebrates out of the mud, and for burrowing into the mud of their pond bottoms, which is how they survive periods of drought.

Surinam Toad (*Pipa pipa*)

This is a most unusual toad from northern South America, in which the females reach a length of 6 inches (15 cm) and males are a little smaller. It is also a purely aquatic species, grayish-brown with pale underparts, with a rectangular shape, flattened body and limbs, and a pointed head. It has large, flipper-like hind feet, but the forelimbs are short and unwebbed, and the fingers end in star-shaped sensory organs instead of the typical frog claws. These organs are highly sensitive and touch is this toad's major sense for locating prey. Surinam toads do not like turbulence, so they seek still waters where they live among the leaves on the muddy bottoms. There is no free-living larval stage in this species, for during the mating process the male helps to attach the eggs to the female's back, where she carries them for the whole incubation period, which may take twenty weeks, and they hatch out as fully metamorphosed toadlets.

Notes

1. Total metamorphosis within the egg occurs in some species, with the young emerging as fully formed froglets, having bypassed the free-living tadpole stage.

2. Unlike in humans, where the ciliary muscle helps to change the shape for focusing (the process known as accommodation).

3. In some adult lizards the pineal eye can be clearly seen as a small spot in the center of the head. It is still functional and acts as a light receptor, which is connected to the pineal gland.

4. Vertical grooves on the sides of newts and salamanders.

5. Two of the three species of endemic New Zealand frogs (*Leiopelma*) still have tiny tails when they hatch as froglets from the eggs, but these are lost as they continue their development on their fathers' backs.

3 The Scent Tasters

Reptiles are believed to have first appeared about 250 million years ago in the Permian Period, and were then the dominant form of life in the following Triassic Period. Intermediate between the lower amphibians and fish and the more recently evolved birds and mammals, they were the first vertebrates to become adapted to a completely terrestrial way of life. They have lungs for breathing air and a protective body covering of scales, shields, or plates instead of the fur or feathers of the other air-breathers, but they vary considerably in form. The tortoises are hard-shelled and mostly high-domed, while some turtles are flat and have soft, leathery shells. Snakes are streamlined and legless, and the typical[1] lizards have four legs and whiptails; then there are the massive, heavily scaled crocodilians. But many are not nocturnal. The nighttime reptiles are mainly concentrated in the order *Squamata* which contains the snakes and lizards. Nocturnal activity in the terrestrial tortoises is practically nonexistent, although some hot desert species like the spurred tortoise (*Geochelone sulcata*) may be crepuscular before they estivate in midsummer. Little is known of the aquatic turtle's or the semiaquatic crocodilian's activities underwater in the dark, and this chapter is consequently devoted to the nocturnal snakes, of which there are many examples, and to the relatively few nocturnal lizards.

There are numerous strictly nocturnal snakes, such as the kraits, coral snakes, and most of the pit vipers; and there are the variable or arrhythmic species like many of the colubrids, boids, cobras, and vipers, which may be nocturnal or diurnal, depending on local conditions (especially the light) or their need for food or warmth. The nocturnal lizards are primarily geckos, of which there are many species, and a few unusual species like the venomous heloderms and the ancient lizard-like tuatara.

■ SNAKES

Snakes are unmistakable, with their elongated, scaly[2] bodies and sinuous movement, forked tongues, smooth earless heads, and staring lidless eyes—their eyelids are fused to form a transparent spectacle over the eye. They belong to the suborder of reptiles known as *Serpentes*, and with their supposed ancestors the lizards form the order *Squamata*. There are about 2,700 species of snakes, wide-ranging animals that occur on all continents except Antarctica, and extend north within the Arctic Circle[3] in Scandinavia and south almost to the tip of South America. Yet they are absent from islands on which lizards still occur, including Ireland and New Zealand. The traditional belief is that snakes evolved from ancient lizards long ago, and lost their legs when they became adapted for burrowing in the soil.

The lethal high-temperature range of snakes is lower than that of lizards, and they are mainly nocturnal, whereas the lizards are primarily diurnal. As with so many nocturnal animals, however, for some snakes there is no strict division between day and night activity. Either their temperature requirements or their feeding needs necessitate activity around the clock, and several species may be seen out and about during the day, especially those that like to bask in the sun, such as the rattlesnakes and rat snakes. Others, like the kraits and coral snakes, are rarely seen in daylight, and numerous others are seasonally nocturnal as they cannot withstand the high temperatures of central continental summers. In the United States the gopher snake (*Pituophis melanoleucus*), common kingsnake (*Lampropeltis getula*), yellow rat snake (*Elaphe obsoleta*), and the copperhead (*Agkistrodon contortrix*), all of which are normally diurnal, become nocturnal during hot summer months. Australia's copperhead, a death adder of the family *Colubridae* (not *Viperidae* like the American *copperhead*), is nocturnal in hot weather and hides in burrows during the day.

Many of the world's most venomous snakes are nocturnal and are responsible for thousands of deaths annually in countries where people traditionally walk barefoot in their villages after dark, and sleep with the doors and windows of their huts open. Coral snakes and kraits on the pathways retaliate if trodden upon; cobras and pit vipers are attracted to habitations in search of rats, and then strike in response to sleepers' movements.

The presence of valuable food resources at night, especially the many nocturnal amphibians and small rodents, makes the dark hours a time of plenty for snakes. In addition to the creatures actively going about their normal business in the dark, all the daytime animals that were expecting a good night's sleep are also there for the taking, if they can be found. It was simply a case of developing the senses to detect them, and the night-active snakes evolved ways to do this.

The Senses

Smell and Taste

The sense of smell is the snake's most highly developed general sense for the recognition of conspecifics and for seeking prey, in both diurnal and nocturnal species, but especially the latter. Like many vertebrates, snakes actually have two senses of smell. One is the standard sense which involves the olfactory system, where nerve endings in the nose perform this function in the normal way. The other is called the vomeronasal system because of the closeness of its sense organ (also called Jacobsen's organ) to the vomer bone in the face and to the nasal bones, and which is associated with their constantly flicking forked tongues. A notch in the upper lip called the rostral groove allows the snake to flick its tongue in and out without opening its mouth. The tongue has no taste buds or any connection to the snake's killing power, however, despite the old belief that they are "stings" in their own right, and is actually a chemical receptor.

The function of the flicking tongue is to collect and transfer scent particles to two pits in the roof of the mouth, from where messages are passed to the brain for analysis. The advantage of a forked tongue is that it allows the snake to determine the direction from which the scent came. These mouth pits contain sensory neurons similar to those in the nose, which detect chemical compounds, including pheromones (which carry messages between animals of the same species) and

Mangrove Snake   *The mangrove snake's continually flicking forked tongue collects chemical compounds (scent particles) from the air and transfers them to pits in the roof of its mouth, from where a message is transmitted to the brain for analysis.*
Photo: Courtesy Harcourt Index

which are connected to the olfactory nerves. This system is used in conjunction with the standard sense of smell, possibly because their rate of breathing does not allow the air to enter their nostrils fast enough to collect and translate scent particles in the normal way. When they use their tongues to "taste the air" in this manner, the pit vipers can follow a hot trail. They do not coil around their prey, but simply strike, inject venom, and then release it, and are then able to track the dying animal, day or night. The organ also alerts the brain to the presence of specific signals that provide information about conspecifics, especially their sex and status.

Vision

Vision is important to diurnal snakes: to the nocturnal snake disturbed while resting or sunning, and to the truly nocturnal ones on bright moonlit nights; but at night sight is less important for snakes than the sense of smell. Experiments to determine the power of the rattlesnake's vision proved they could detect movement up to a distance of 15 feet (4.5 m), although they picked up vibrations from the ground long before this point was reached. Although long-distance vision is of no great value to a snake crawling through the grass at ground level, it is certainly useful for arboreal species, which hang from a branch poised to seize a passing bird. The snake's field of vision is mainly monocular, with a slight overlap; the greatest degree of binocular[4] vision occurs in the tree snakes where more accuracy is needed to secure prey, as a second chance at their victim is rarely possible.

Unlike the nocturnal mammals and birds, night-hunting snakes do not necessarily have greatly enlarged pupils; in fact the eyes of many are no larger than those of diurnal snakes, but they have larger lens apertures and cornea, which improves their light-gathering ability but reduces visual acuity (sharpness of vision). Pupil shape varies, and in most nocturnal species is vertical in contraction. Most colubrid snakes, at least the nonvenomous ones, have round pupils, while the venomous species such as the brown tree snake (*Boiga irregularis*), night snake (*Hypsiglena torquata*), and cat snake (*Telescopus fallax*) have vertical pupils. The highly venomous southern vine snake (*Thelotornis capensis*) of southern Africa has long, keyhole-shaped pupils; and the long-nosed whip snake (*Ahaetulla nasuta*) of Southeast Asia has a very elongated head and long eyes with horizontal slit pupils in line with the shape of its head. Binocular vision is most pronounced in these species. Snakes, just like the mammals and birds, have rods and cones in their retinas and are therefore assumed to have some degree of color vision in daylight, but their photoreceptors lack colored oil droplets, and they are therefore unlikely to have such broad-ranged color perception as the higher vertebrates.

Hearing

Snakes lack the well-developed communication systems of the mammals and birds, and are relatively silent animals—a low hiss being their characteristic sound. They have no external ear, no eardrums and no middle ear; traditionally, their sense of hearing was believed to be just a sensitivity to ground vibrations, and they were thought to be deaf to airborne sounds. However, recent research has shown that the snake's tiny columella bone (the equivalent of the stapes bone in the mammalian ear) connects the jawbone to the inner-ear canals instead of to the eardrum as in other reptiles. Sounds picked up through the skin as vibrations are passed along it to the jawbone, and then transmitted through the bones of the jaw and skull to the cochlea in the inner ear, which produces electrical responses that are passed to the brain.

Touch

The sense of touch is a highly developed sense in snakes, which is understandable for a creature that spends most of its life on its belly in contact with the substrate or branches. Their bodies are covered with tactile receptors, which enable them to "feel" the slightest change in the surface they are crossing, allowing them to change the muscular movements needed for locomotion. Touch is also the snake's most important sense during encounters with its conspecifics. Males have ritual combat "dances" in which they intertwine like wrestlers, oblivious to the happenings around them. The actual mating encounters between male and female often involve a writhing and entwining courtship which may last for an hour before mating occurs, but the most unusual mating encounters occur in the garter snakes when they emerge from hibernation. Many males seek a receptive female, even following her into the branches of low bushes, where they intertwine in a ball while attempting to mate, and then separate when one male is successful.

Heat Sensing

Many snakes have an extra sensory system, with which they are able to locate prey by heat radiation. In the pythons and boas heat-sensing pits are situated along the edges of the mouth. The pit vipers have a more sophisticated system, often called the "sixth sense," which involves deep facial pits containing nerves and receptor organs that can detect temperature variations of a fraction of one degree. These are both discussed in more detail under the appropriate snakes.

Hiding by Day

Hiding is the main preoccupation of most nocturnal snakes during the day, from the sun and from potential predators. Avoiding the sun is particularly important for the strictly nocturnal species such as the coral snakes, kraits, and pit vipers, which cannot abide exposure to light. Wagler's pit viper (*Tremeresurus wagleri*), the snake that can be seen in large numbers in the snake temple on Malaysia's Penang Island, becomes sluggish when exposed to the sun's rays. Direct sunlight, on a hot day, has such a tranquilizing effect on the krait that it is unable to bite, and excessive exposure is fatal due to the increase of body temperature. To escape the sun, death adders and coral snakes burrow, but the less strictly nocturnal species such as the cobras, adders, and many of the colubrids may be seen sunbathing for short periods to raise their body temperature. Within the Arctic Circle the adder must sun bathe to raise its body temperature sufficiently to become active in the cool, short summers there.

All snakes are predators, whether of insects or small mammals or a whole host of creatures in between, but despite being killers in their own right, they are also themselves fair game to a wide range of animals. Nocturnal activity certainly does not provide protection from predators, and the night-active snakes, even highly venomous species, fall prey to owls, raccoons, mongooses, and native cats, and even to their own kind. Many snakes are cannibalistic and just eat snakes—other species and their conspecifics. King cobras, kraits, king snakes, and milk snakes are all snake-eaters, which must be housed separately by zoos and herpetologists. Even their newly hatched young cannot be kept together or with other snakes after they have shed their skin for the first time and are then looking for a meal. Rattlesnakes in North America have been killed by their fellow pit vipers the moccasins and even by the nonvenomous king snake. In South America the role of cannibal snake is taken over by the powerful mussurana (*Clelia clelia*), a rear-fanged colubrid snake that is a major predator of the plentiful pit vipers of the genus *Bothrops* (of which sixteen species live in Brazil alone), which it kills by constriction.

The nocturnal snakes protect themselves during the day by hiding in tree holes, in deep leaf litter, in other animals' burrows, or by burrowing into the soil themselves. But they are not completely safe, and are sniffed out and eaten by wild pigs, coatimundies, armadillos, and monitor lizards; and if they do sunbathe there is a host of birds looking for them, including giant kingfishers, storks, and secretary birds, and even eagles, which specialize in catching snakes. Some just "hide in full sight," simply relying on their camouflage to protect them. Surprisingly, the young of some snakes appear startlingly colored, totally different from their parents (see color insert), although this obviously has cryptic value in their habitat.

Some of the Species

Boas and Pythons

The boas and pythons are generally linked together in the family *Boidae*, and are usually just called constrictors, as they kill their prey by coiling around its body and suffocating it. The family is then subdivided into the boas (*Boinae*), and the pythons (*Pythoninae*). There are several differences between them, including such minor anatomical details as the boa's lack of teeth on the premaxilliary or suborbital bone of the skull, but the major difference lies in their method of reproduction, for boas give birth to live young, whereas pythons lay eggs. The constrictors usually have two lungs, unlike the general arrangement in snakes, in which the left lung is reduced and sometimes absent and the right one is enlarged and lengthened. Most species have external "spurs," that are vestigial hindlimbs on either side of the vent, which are quite large in the pythons.

The most impressive aspect of these snakes is their large size and their constricting powers—both of which have been exaggerated by fiction and adventure writers, and even eyewitnesses. The reticulated python and the anaconda share the title of the world's largest snake, reaching a length of about 33 feet (10 m). But despite tales of snakes measuring 60 feet (18.3 m), carcasses or skeletons authenticating such lengths have never been produced, and it is well known that snake skins can be stretched after removal from the body. The next-largest species is the African python which may reach 25 feet (7.6 m) in length, and then the Indian python and the Australian amethystine or scrub python, both of which may reach a length of about 20 feet (6 m), although such long individuals in all these species are rare. The boa constrictor, or common boa, often popularly thought to be the largest snake, is actually the smallest of the giant snakes, reaching a length of only 13 feet (4 m).

The constrictors can swallow large animals, but their abilities have also been greatly exaggerated. Snakes have been recorded as swallowing large oxen, which is impossible; even the largest anaconda could only just manage to swallow an average-sized person. The largest natural prey of the giant contrictors are wild pigs, peccaries, young tapirs, adult capybara, small deer and antelope (at least antler- and horn-less females), and even crocodilians up to 6 feet (1.8 m) long. In addition to their highly mobile "elastic" jaws, their windpipe opening can be extended so they can breathe while there is a large object obstructing the throat. Seizing from ambush on the ground is the typical hunting behavior of the large sedentary snakes. They do not chase prey like the racer or the king cobra, and rarely hang from trees waiting to grab passing explorers as the early travel books depicted.

Some boas, especially members of the genus *Corallus*, and many of the pythons, have an extra sensory system, although not quite as sophisticated as the "sixth sense" of the pit vipers. They have heat sensing pits located in crevices between the scales on the mouth (see color insert), called supralabial pits in the top jaw and infralabial pits when they are in the bottom jaw, which sense the heat given off by their warm-blooded prey, and are sensitive to temperature variations of as little as 0.18°F (0.10°C). The value of this sensory apparatus is shown to remarkable effect in the darkness of caves by the Cuban boa, that reaches a length of 12 feet (3.6 m) and which locates and seizes bats, even on the wing.

Common Boa (*Boa constrictor*)

This snake has a very wide range from Mexico through the tropics to southern Brazil and also occurs on several islands in the Lesser Antilles. There are many races, ranging in size from the Crawl Key boa at 2 feet (61 cm) long to the Amazonian boa, which may reach 13 feet (4 m) in length and weigh 80 pounds (36 kg); but like the anacondas and reticulated pythons, much larger specimens have been occasionally reported. As a species the common boa is the most imported and probably the most frequently bred snake, with wild-caught and captive-raised specimens regularly available. The Amazonian boa (*Boa c. constrictor*), has several geographic variations with different body markings, the most common ones being those known to the pet trade as the Brazilian and Peruvian red-tailed boas. Another favorite subspecies as a pet is the Colombian boa (*B. c. imperator*), one of the smaller races which is usually good-tempered. The Argentine boa (*B. c. occidentalis*) is a rather dark form, and the Dominican or clouded boa (*B. c. nebulosus*) has pale markings on a grayish-brown background.

Emerald Tree Boa (*Corallus caninus*)

This very attractive arboreal boa reaches a length of 5 feet (1.5 m). Two types are recognized, although they have yet to receive subspecific status—the northern emerald boa from the Guiana Shield and the Amazon emerald boa, which occurs south and west of the Guyana Highlands. Both have yellow undersides, but are otherwise quite distinct; the northern one is lime-green with separated transverse white dorsal markings, whereas the Amazon type is dark emerald-green and its white markings are connected along the spine. Their scalation is also different, the northern snake having much larger scales on the head than the southern one; and they also differ in temperament—the northern boa being generally being more aggressive. In total contrast to their green parents, most young emerald boas are reddish-brown with white markings, although some northern babies lack the markings and occasionally are green at birth. When resting or basking they characteristically drape over a branch, often close to the ground.

Green Anaconda (*Eunectes murinus*)

The anaconda is probably the world's largest snake, with more records of specimens reaching 33 feet (10 m) in length than its close rival the reticulated python. It is certainly heavier than the python, a massive snake that can have a diameter of 15 inches (38 cm) and may weigh in excess of 400 pounds (181 kg). Semiaquatic by nature, the anaconda has a large head, thick neck, and a very muscular body. Its eyes and nostrils are on top of its head so it can be almost totally submerged while waiting in ambush for capybara, peccaries, pacas, deer, fish, and crocodilians. Providing water is nearby, the anaconda can be found in rain forest, deciduous forest, savannah, and seasonally flooded grasslands. Females in breeding condition emit pheromones which are carried on the wind to males who can track them by picking up the scent on their tongues. The green anaconda is a native of northern South America, including the Orinoco and Amazon River basins, plus the island of Trinidad. It is dark green with black spots on its back and sides, the latter having yellow centers. Another species, the yellow anaconda (*Eunectes notaeus*), lives to the south of the green anaconda's range, in southern Brazil, Uruguay, and northern Argentina.

Reticulated Python (*Python reticulatus*)

A length of 33 feet (10 m) has been recorded for this species, which therefore shares with the anaconda the title of the world's longest snake, but it is not as heavy, the maximum recorded weight being just over 300 pounds (136 kg). It is a tropical forest animal, with a range from southern Burma through the Malay Peninsula into Indonesia and the Philippines. It has a reputtation for being bad-tempered, and together with its potential for huge size, is not as popular a pet as the smaller Burmese python. Its name derives from its distinctive patterning, a kaleidoscope of colors including blotches of tan, brown, dark-gray, and gold outlined in black on a background of silver-gray. Like the anaconda its prey has included humans, and after a large meal it can rest and digest for several months. A captive animal is recorded eating a pig that weighed 120 pounds (54 kg), and another zoo python did not eat for almost two years after a large meal. Unlike the boas they are egg-layers and have the unusual habit, for a reptile, of actually incubating their eggs, coiling around them and raising their body temperature a few degrees through muscular contractions.

Diamond Python (*Morelia s. spilotes*)

The diamond python occurs east of Australia's Great Dividing Range in Victoria and New South Wales, where it lives in dry, rocky country and is accustomed to cool winters with temperatures that may drop below freezing. At such times it shelters in tree hollows, leaf litter, or in rock crevices, which is where it also spends the daylight hours. It is one of the many races of the carpet python (*Morelia spilotes*), which is widespread throughout Australia and New Guinea. The races vary in maximum length, but the largest can reach 14 feet (4.3 m), and although they are basically terrestrial they are good climbers. The diamond python is a stocky snake with a broad head, and is black with creamy-yellow spots on its back; it has a cream belly with gray blotches, and its markings are sometimes clustered in the shape of diamonds. Its diet consists of rodents, small marsupials, and birds, whose capture is aided by the infrared sensing pits in its labial or lip scales.

"Typical" Snakes

This is the largest and most widespread group of snakes, containing almost 80 percent of all living species, and forming an even larger percentage of the snakes of some regions, especially North America. For want of a common family name, and with such a great diversity of members that do not seem to fit anywhere else, they are lumped together in the family *Colubridae*, and have been called typical snakes, or just colubrids. They are distributed throughout the world, except the polar zones, New Zealand, and Ireland, and both southern tips of Australia. In fact they are the dominant snakes on all continents except Australia, where the elapids dominate. They are just as wide-ranging in their habitat and behavior, and show great variation in color and pattern. Many of the species are nocturnal, and most are terrestrial.

The family *Colubridae* contains mostly nonvenomous forms (*aglyphs*), which have simple teeth, and mildly venomous species (*opisthoglyphs*), which have a pair of fangs located at the back of the upper jaw and are therefore called "rear-fanged" snakes. Their fangs are not hollow, however, but grooved to allow the venom to run down, so the snake must chew on its prey for the venom to penetrate the wound. With the exception of the highly venomous arboreal species from southern Africa—the boomslang (*Dipholidus typus*) and two vine snakes of the genus *Thelotornis*—the colubrids are rarely harmful to large animals or to man. They bear no traces of hindlimbs or pelvic girdles, which the "primitive" snakes have, and are therefore also often called "advanced" snakes. Their left lung is either very small or absent, and they have characteristic large, strap-like scales.

Ring-necked Snake (*Diadophus punctatus*)

The nonvenomous ring-necked snake lives in moist areas in woodland and grasslands, farms, and gardens in North America from southern Canada to Mexico. Across this large region eight subspecies are recognized. In the more arid parts of its range, it stays above 2,400 feet (732 m) near water courses. It is a slender terrestrial snake that hides beneath rocks and rotting logs during the day, emerging at night to look for its preferred cold-blooded prey, which includes salamanders, frogs, lizards, and worms. The ring-necked snake grows to a maximum length of 30 inches (75 cm) and has a dark-olive or almost black body with a dark head and a bright-yellow or orange ring around the neck. Its orange belly is usually spotted with black, and the undersurface of its tail is bright red; when alarmed it coils its tail to show the underside. It lays its eggs in early summer, often in communal nests used by several females, in damp places such as piles of decaying leaves that are exposed to the sun.

Corn Snake (*Elaphe g. guttata*)

The corn snake, also known as the red rat snake, is the most colorful of the New World rat snakes, as it is generally marked with reddish blotches bordered with black on a pale-yellowish background. It is one of the smaller species of rat snakes, averaging 3 feet (92 cm) in length, although individuals over 6 feet (1.8 m) have been recorded. It has a wide distribution across the southern United States and northern Mexico, where it prefers drier habitat such as dry river bottoms, rocky hillsides, and pine barrens. Like the other rat snakes it is nonvenomous, and although mainly terrestrial, is a good climber, its weakly keeled ventral plates allowing it to scale vertical tree trunks with rough bark by pressing its body against the bark and hitching upward with its keeled plates. The corn snake is the most popular of all snakes kept as pets, and has responded to improved breeding techniques by producing some very attractive color mutations since the first albino was bred in captivity in 1962. Each year U.S. breeders produce thousands of amelanistic corn snakes in which the lack of normal black pigmentation results in numerous mutations ranging from blood-red to white.

Mangrove Snake (*Boiga dendrophila*)

The mangrove snake is glossy black with thin yellow bands, which in some individuals do not meet on top of the back or on the stomach. The bottom half of the head and the throat may also be bright yellow. It is one of the tropical cat snakes, so named for its vertically slit cat-like pupils in bright light. It is crepuscular and nocturnal and rarely leaves the trees, even laying its eggs in arboreal termitaries or in tree hollows. It sunbathes during the day, stretched out across the foliage, and can hang from branches by the lower one-third of its body, leaving the rest free to strike at prey, which includes a variety of arboreal creatures such as amphibians, lizards, other snakes, and birds and their eggs and nestlings. The mangrove snake is a native of Malaysia and Indonesia, usually in wet areas and especially in mangrove swamps. Rear-fanged and mildly venomous, it is a very long and slender species with a body that is flattened laterally and can reach 9 feet (2.7 m) in length.

Brown Tree Snake (*Boiga irregularis*)

One of the mildly venomous colubrids with vertically elliptical pupils, the brown tree snake has rear fangs that have open grooves on their sides for the transport of venom, and therefore chews on its victim to ensure venom penetration. Although its bite causes discomfort in humans it has never been fatal. It is a native of the coastlands and hinterland of Australia, from northern Western Australia to New South Wales; plus the islands of eastern Indonesia, Papua New Guinea, and the Solomon Islands. Mainly a frog and lizard eater, it is the snake that has seriously damaged the fauna of Guam, after being accidentally introduced onto the island after World War II. It has since destroyed the populations of several ground-nesting birds, including the flightless Guam rail. Sniffer dogs are now employed to check air cargo to prevent the snakes from being transferred to other islands.

Brown tree snakes reach a maximum length of about 5 feet (1.5 m), and vary considerably in pattern and color. They may be pale brown, banded pale brown and dark brown or brown and black, and are strictly nocturnal, but may bask in crevices or similar out-of-sight places. They congregate with others of their kind, and sometimes other species of snakes, during the winter months in southern Australia.

Elapids

Cobras are members of the family *Elapidae*, which contains a number of other equally dangerous snakes, many of which are nocturnal, including the coral snakes, kraits, and the death adder in addition to several nocturnal cobras. The

Brown Tree Snake *A native of eastern Australia and islands to the immediate north, this snake was accidentally introduced to Guam shortly after World War II, and has since been responsible for the extinction of several of the island's endemic birds, which had evolved in the absence of such predators.*
Photo: Courtesy USFWS, Gordon H. Rodda

elapids have a pair of fixed and immovable venom fangs in the front of the upper jaw, which are hollow for the transport of venom. Their venom is highly toxic, and the family contains some of the world's most venomous species, which are responsible for more human deaths annually than any other group. Their teeth fit into grooved slots on the floor of the mouth when it is closed, and a muscle presses down on the venom gland as the snake bites, forcing venom into the wound. Most of the elapids are hunters, like the diurnal mambas and the cobras, which actively seek and chase their prey.

Elapids are the dominant snakes in Australia, where about seventy-five species live, many of them very dangerous animals. In the absence of competition when Australia broke away from Gondwanaland, the ancestors of the present-day snakes adapted to the many vacant niches and evolved into many distinctive species, a case of adaptive radiation—the diversification of animals from a common ancestor. Australia is the only country where venomous snakes outnumber the nonvenomous ones. The cobras are the most recognizable elapids, as they can depress their necks horizontally to form a hood or cape, created by the extension of the ribs behind the head. The largest venomous snake, although not nocturnal, is the king cobra, which can reach a length of 18 feet (5.5 m).

Common Krait (*Bungarus caeruleus*)

The krait is a highly venomous nocturnal snake of India, Sri Lanka, and Pakistan, which normally preys on lizards, frogs, and rats, but is frequently cannibalistic. It searches for rats and other snakes at night in the vicinity of habitations, and relies on sight and vibrations when hunting. As it has short fangs it must chew its victim to inject venom into the wound, but its venom is highly toxic to humans, and kills many people in India annually, often as a response to sleepers' movements as it prowls through their dwellings at night, or as a result of being stepped upon by a bare foot. Farmers in the drier districts are particularly vulnerable when they sleep in open huts in the fields. The krait is considered one of the four most dangerous snakes in India, the others being another elapid—the Indian cobra—and two vipers—the saw-scaled viper and Russell's viper. The common krait reaches a length of 6 feet (1.8 m) and is slate-colored with thin white bands around its body that fade toward the tail.

Western Coral Snake (*Micruroides euryxanthus*)

A small, brightly banded snake, just 22 inches (56 cm) long and the thickness of a pencil, the coral snake has alternating wide red-and-black bands and narrow yellow-and-white rings encircling its body. It lives only in the Sonoran Desert of Arizona and northern Mexico and the southwest corner of New Mexico; it prefers upland desert regions and dry rocky areas where the saguaro cactus grows. The western coral snake is nocturnal and spends most of its time under rocks or buried in the sandy soil, coming out after the spring rains to search for lizards and snakes; it paralyzes them with its neurotoxic venom, which is more powerful than a rattlesnake's. To warn off predators it hides its head, raises its tail to expose the underside and then makes a popping noise by everting the lining of its cloaca. It lays two or three eggs each year in late summer.

Egyptian Cobra or Asp (*Naja haje*)

The Egyptian cobra is another dangerous snake that is responsible for many deaths annually across North Africa, where it lives in woodland, grasslands, and semidesert. It is a truly nocturnal snake, which can move fast when searching for toads, lizards, and roosting birds, and is very aggressive toward people, attacking repeatedly when it feels threatened. The asp was worshipped in Ancient Egypt and incorporated into Pharaonic crowns, and Cleopatra assumed she would become immortal when she called for an asp to end her life, according to legend. Asp venom has been used in medical research to destroy the membranes surrounding viruses to control them. The asp reaches a maximum length of 8 feet (2.4 m), but most specimens are about 6 feet (1.8 m) long. It has a small, flat head and rounded snout, and varies in color from grayish-yellow to dark-brown. Its hood has dark markings on the front but the back is unmarked.

Chinese Cobra (*Naja naja atra*)

One of the ten subspecies of the Indian or Asiatic cobra (*Naja naja*), the Chinese cobra has a wide range from the warmer southern latitudes of China through Vietnam and Cambodia to Thailand, and also occurs on Hong Kong. Several color mutations have been described in this very large range, but the typical form is a plain snake, its body coloring generally being a yellowish-tan with the top of its head dark brown. The large hood is marked on the back with an irregular off-white patch with brown spots. The Chinese cobra is a thick-bodied snake, largely terrestrial in its habits (although it does occasionally climb trees), and it has very potent neu-rotoxic venom with which it immobilizes its prey of small mammals, birds, and lizards. It is shorter than most other cobras, with an average length of 5 feet (1.5 m).

Vipers

This family of snakes, the *Viperidae*, contains a number of nocturnal species, which are divided into two subfamilies to separate the vipers or adders (*Viperinae*) from the quite different pit vipers (*Crotalinae*). The vipers include the adder, Britain's only venomous snake, and the very stout-bodied vipers of Africa—the rhinoceros viper, puff adder, and Gaboon viper. The pit vipers include the more familiar rattlesnakes, copperhead, and water moccasin or cottonmouth, plus many arboreal Asiatic pit vipers and the largest New World venomous snake, the bushmaster. They have heat-sensing pits and are considered more advanced than the *Viperinae* snakes.

The members of the family *Viperidae* are the only snakes with hollow, hinged fangs situated at the front of the upper jawbone. In their erect and usable position, the fangs are too long for the snake to close its mouth, so they fold back against the roof of the mouth when not in use. Their fangs can therefore grow quite large—up to 2 inches (5 cm) long in a mature Gaboon viper—and venom from their glands can be injected quickly by the large muscles at the back of the head, which results in their characteristic triangular shape, accentuated by the thin neck. When the snake strikes, its fangs are erected by a special bone which connects with the jawbone and acts as a lever. The night adders (*Causus* spp), which live in sub-Saharan Africa, are an exception, differing from the typical adders in the elongated shape of their heads, as the venom sacs extend backward into the body.

These snakes are ambush predators, which strike and then follow the scent of their stricken prey. They are generally sedentary and well camouflaged, and just lie quietly in wait until something edible comes within reach. They can tell from the intensity of the vibrations whether the approaching animal is too large for them, and they will lie still or slip quietly away. After striking their prey and injecting venom they trail it by means of their Jacobsen's or vomeronasal organs until it succumbs, which is just a few seconds for a small rodent as the hemotoxic venom is extremely potent and acts very quickly. The viper's jaws are more loosely jointed to the skull than those of other snakes so they are able to open wider and swallow larger prey. Their large meals result in a correspondingly longer digestion time, giving the impression that these snakes always seem to be resting. The vipers occur throughout the world except Antarctica, Madagascar, Australia, and New Zealand.

Puff Adder (*Bitis arietans*)

A widespread species of sub-Saharan Africa's savannah and bushveldt and now of the cultivated lands, it is one of the most feared snakes on the continent, as it comes into increasing contact with humans as settlement, agriculture, and tourism invade its domain. It is now the most dangerous snake in Africa due to its presence in populated areas. Its lethal qualities are increased by large venom glands and the long, deeply penetrating fangs that inject cytotoxic venom, which destroys cells. The puff adder is a rough-scaled, heavy-bodied snake that reaches 5 feet (1.5 m) in length, and has large nostrils that point upward. It is a slow-moving animal, always very bad-tempered, which hisses and puffs up and strikes out in all directions when annoyed. Yellow-brown to light-brown in color, it has black pale-edged chevron markings on the back which may be almost lost in the general speckling; and its belly is white or yellow. It is perfectly camouflaged in its dry, grassy, and rocky habitat, which greatly increases its danger to humans. But its lethal bite does not provide total security, for it is preyed upon by warthogs, honey badgers, and several raptors. Its own diet includes small mammals, especially rats and mice, and ground-dwelling birds.

European Adder or Viper (*Vipera berus*)

The most widespread venomous snake in Europe, the adder occurs in all the countries except Ireland, and is the only dangerous species in Britain, although its bite has rarely been fatal to humans. Its range also extends westward across northern Asia to the Pacific Ocean, and it has the most northerly distribution of all snakes, living just inside the Arctic Circle in Siberia. The male adder's body color is creamy-yellow to gray and the female is reddish-brown; both have a dark zig-zag pattern along the spine, and are usually about 24 inches (60 cm) long. Throughout its range the adder hibernates during the winter months, but despite being mainly nocturnal it must bask frequently to warm up, especially during the weak sunshine of spring and autumn. In the far north the short summers restrict its active season to a few months, and it hibernates in deep crevices and sink-holes to stay below the frost, which penetrates at least 6 feet (1.8 m). The short summers in the far north are very hard on the female adders, which must mate, complete the internal incubation of their eggs, and give birth to live young, and then replace their reserves for the winter dormancy with their chief prey—rodents, frogs, and toads. Their young, born late in summer, must also find sufficient food in a few weeks to store energy for their first winter, but many do not survive.

Pit Vipers

Pit vipers (of the subfamily *Crotalinae*) are the most specialized of all nocturnal snakes, adapted for hunting in total darkness. They are widespread throughout the warmer regions of the world, except Africa, and are ecologically adapted for life in a wide range of habitat. There are terrestrial rain forest pit vipers, such as the 10-foot-(3 m)-long bushmaster (*Lachesis muta*); the prehensile-tailed arboreal species, of which the 4-foot-(1.2 m)-long Nicobar Island pit viper (*Trimeresurus labialis*) is the largest; desert dwellers like the sidewinder (*Crotalus cerastes*); and the semi-aquatic moccasin (*Agkistrodon piscivorus*). They occur from sea level like the fer de lance (*Bothrops atrox*) to 16,000 feet (4,880 m) in the Himalayas, where *Agkistrodon himalayanus* lives. Most species are live bearers, with the bushmaster being the only New World pit viper to lay eggs, while in the Old World, the Malayan pit viper (*Agkistrodon rhodostoma*) is also an egg-layer.

Rattlesnakes are the most numerous pit vipers in the New World, occurring from southern Canada to Argentina. Composed of horny, loosely interlocking segments at the end of the tail, the rattle functions as a warning only. Newborn rattlers have a single button on their tails, and add a segment every time they shed their skins, which is determined by their growth rate. This occurs faster in young snakes, which may add up to four rattle segments annually, slowing down to one per year as they mature. Rattlesnakes have colonized a wide range of terrestrial habitats, from grassland and desert to dense woodland.

Pit vipers are highly dangerous snakes that kill many people around the world annually. Their sinister appearance heightens the general apprehension of snakes and hostility toward them. In common with other species, they have the typical—"cold" expressionless eyes and they lack eyelids, but their sinister looks are enhanced by their vertical pupils, the enlarged scale that hangs over the eye, and their large, triangular-shaped heads. They have almost unequalled killing power, with long, deeply penetrating fangs and large amounts of venom, up to 600 mg being produced by a large eastern diamondback rattlesnake at one "milking." Rattlesnake venom retains its qualities for long periods and is even dangerous when fifty years old. It breaks down the walls of the blood vessels and the leucocytes, but there are specific variations. The South American, pygmy, and western Mexican rattlesnakes are especially dangerous as their venom also contains a neurotoxic element that acts on the nervous system, so a special antivenin is needed to treat their bites. The digestive enzymes present in the venom of the eastern diamondback rattlesnake have a necrotizing effect on flesh, and its victims have lost limbs. Fer de lance venom has a very spectacular effect, causing hemorrhaging, with blood oozing out of the body's orifices.

Like the *Boidae*, which have heat sensors on their lips, the *Crotalinae* also have a heat-sensing system, but theirs is a far more sophisticated one, which sends a "heat image" to the brain. They have a sensory loreal or facial pit, about ¼ inch (0.5 cm) deep, below and in front of the eye on each side of the head, almost forming a triangle with the eye and the nostril. The pit is a cavity in the maxilliary bone, covered with a membrane and connected to the outside by a tiny hole. The membrane is supplied with about 7,000 nerve endings and receptor organs sensitive to the infrared radiation (invisible rays beyond the red end of the spectrum) whose wavelengths are too long to stimulate the human retina, and too short to produce the stimulation of heat in our receptors. They can detect heat emanating from their prey, being sensitive to temperature differences as low as 0.02°C, and can therefore accurately locate prey animals. On a night when the ambient temperature is lower than that of a rat kangaroo, a rattlesnake can detect the rodent 18 inches (46 cm) away by its body temperature. It has been proved that the pit is a true sense organ, which other animals do not possess. Researchers blindfolded rattlesnakes, removed their tongues and plugged their nostrils and discovered they were able to strike accurately at covered light bulbs, confirming that their pits guided the snakes to the source of heat.

Copperhead   *A pit viper and therefore a relative of the rattlesnakes, the copperhead has a heat-sensing pit on each side of its head in front of its eye. This pit is richly supplied with nerve endings and receptors sensitive to infrared rays, which can detect the heat radiating from its prey, allowing it to hunt in total darkness.*
Photo: Courtesy USFWS

Western Diamondback Rattlesnake (*Crotalus atrox*)

This species is the largest "rattler" in western North America, occurring in the southwestern United States and adjoining northern Mexico and growing to about 7 feet (2.1 m) long. It is a thick-bodied snake with keeled scales, and has the triangular head and narrow neck typical of the vipers. The diamondback varies from gray or yellow to dark brown, and has light-brown to black diamond-shaped blotches that fade toward the tail. The tail is ringed with black and white and tipped with the segmented rattle to which a new segment is added each time the skin is shed. It is a stubborn snake that usually stands its ground when threatened, but should be left strictly alone, as its bite is potentially deadly unless medical treatment is immediately available. A nocturnal snake, it is active by day in the coolness of spring, and basks in the late afternoon. Largely a rodent-eater, it preys upon kangaroo rats, rabbits, and ground squirrels whose burrows it can enter at night. It is viviparous and gives birth to about twenty young in late summer.

Bushmaster (*Lachesis muta*)

The bushmaster holds two records. It is the longest venomous snake in the New World, and the longest viper in the world. It is a forest snake, especially of the cooler cloud forest, a few thousand feet above sea level, in Central America and the northern half of South America, plus the island of Trinidad. There is also an isolated population in the coastal forests of eastern Brazil. The bushmaster's longest recorded body length was almost 12 feet (3.6 m), but such huge specimens are rare. It has large fangs that may reach 1¼ inches (3 cm) in length and its glands produce an enormous amount of venom, with an average yield of over 400 mg of dried venom, compared to the copperhead's yield of 50 mg. The bushmaster's background coloration is reddish-brown to pinkish-gray, covered with a diamond pattern. Unlike most pit vipers it lays eggs, up to twelve per clutch, usually in a burrow dug by a small mammal such as a paca or armadillo, and aggressively guards the eggs until they hatch.

Wagler's Pit Viper (*Tropidolaemus (=Tremeresurus) wagleri*)

A widespread snake in Southeast Asia, including the Malay Peninsula, the Philippines, and most of the Indonesian islands, Wagler's pit viper is probably the most well known of the Asiatic pit vipers due to its presence in large numbers in Penang's snake temple, where it supposedly brings good luck to the worshippers. Unlike most of its kind, it is a mild-mannered snake that does not generally bite humans, and its bite is not considered deadly, but the snakes that temple visitors handle have been de-fanged just to be safe. Wagler's pit viper is a snake of the lowland and submontane rain forest and mangrove swamps, where it spends the day coiled around a branch in the shade, as it cannot abide exposure to the sun. When mature it measures just under 3 feet (92 cm) in length. There is the usual variation in color befitting an animal with such a wide range, especially in the isolated insular forms, but the temple snakes have a black background with yellow bands across the black and green spots on the black between the bands, and their bellies are usually yellow. Like most of the pit vipers they are live bearers, and produce up to fifteen young at a time.

■ LIZARDS

Lizards belong to the suborder *Lacertilia*, which was until recently called *Sauria*, and together with the snakes they form the order *Squamata*, the largest order of reptiles. There are about 3,000 species of lizards, and they are considered the most successful of the modern reptiles, with a wide distribution on all the continents except Antarctica, and on many oceanic islands which they reached accidentally, most likely on driftwood. They have adapted to survive in a variety of habitats, including the rain forest, grasslands, desert, and swamps.

Lizards are generally distinguished from snakes by their limbs, which most species have, but there are actually many "legless" forms in which their vestigial legs are not obvious externally, and therefore they resemble snakes. The most obvious difference between all lizards and snakes is the presence of eyelids in the lizards, and a few species even have a transparent bottom lid so they can still see with their eyes closed. Also, together with the crocodilians, turtles, and the amphibians, the lizards that have closable lids have the extra clear eyelid known as the nictitating membrane or third eyelid, which moves horizontally across the eye and protects it while still allowing good vision.

Nocturnal behavior is far less common in lizards than in snakes, and in only one group—the large family *Gekkonidae*—are there many night-active species. In fact, far more geckos are nocturnal than diurnal, and they make up the bulk of the nocturnal lizards. Although the division between nocturnal and diurnal habits is well defined in lizards, there are a number of species that are normally active by day in cool weather, but become nocturnal in their habits when it is very hot. There are several examples of such seasonal nocturnal habits in the lizards of Australia's hot "Red Centre," such as the centralian blue tongue (*Tiliqua multifasciata*) and the pink-tongued skink (*Tiliqua gerradii*). The desert regions of the southwest United States and neighboring Mexico probably have the greatest concentration of nocturnal lizards, with numerous geckos and both species of venomous heloderms.

Some of the Species

Geckos

The geckos are mostly small, soft-skinned lizards, which range in size from some of the tiniest lizard species less than 2 inches (5 cm) long to others that are 16 inches (40 cm) in total body length. Their family *Gekkonidae* is a very large one, containing about 700 species, 75 percent of which are nocturnal. With few exceptions, the nocturnal species are not brightly colored, being clad mainly in browns and grays. They hide by day in crevices, attics, beneath bark, and in any dark, undisturbed place. In contrast the geckos which are active during daylight, appropriately called day geckos, are bright green with red markings. Geckos live on tree trunks in the rain forest, among the rocks in semiarid regions, and are familiar in houses in warm climates where they benefit from insects attracted by the artificial lighting; and where their ability to climb glass and smooth walls is so obvious. In this situation they are the most well known of all the nocturnal lizards.

Moorish Gecko   *A tiny hatchling of this Mediterranean species that reaches a length of 6 inches (15 cm) when adult. A lizard of the arid rocky regions, it has been introduced into California where it is now established.*
Photo: Courtesy Paolo Mazzei

The structure of their feet allows many geckos to scale smooth surfaces, even glass. Most species have sharp claws, and the undersides of their toes have pads covered with microscopic filaments (called setae), which look like stiff hairs, and interaction between these and the surface create an intermolecular force that holds the gecko to the surface, allowing it to walk up a vertical sheet of glass. However, geckos' feet vary according to their habits and some lack glass-climbing toes. The bent-toed gecko (*Cyrtodactylus peguensis*) of Malaysia is one of these, with toes that are permanently turned upward at their base and downward at their tips, so that its claws are held at right angles, allowing it to climb trees with ease. The banded gecko (*Coleonyx variegatus*) of the American desert regions is mainly terrestrial and also lacks toe pads and uses its claws to climb over rocks in search of insects; the members of the genus *Eublepharis*, which includes the leopard gecko, also lack digital setae.

Sight is the most effective sense of the nocturnal geckos. Most have large eyes, with larger pupils, lens apertures, and corneas, which improves their ability to gather light, although their visual acuity is reduced. Some also have parietal eyes—the photosensitive organs on top of diurnal lizards' heads—which are believed to be absent in most nocturnal lizards; but the lizard-like tuatara has a well-developed one, albeit covered with scales. In lizards it is believed to act as a light sensor, and to be involved with circadian rhythms, and is therefore important in the reproduction process. It is also active in thermoregulation or temperature control and likely regulates basking time.

Like the nocturnal frogs and toads, but unlike the snakes which rely more upon scent-tasting, geckos depend upon sight to hunt at night. They are all insectivorous or carnivorous and are stimulated by the movement of their prey. They lack the snake's constantly flicking forked tongue, and although they have vomeronasal organs, these are probably not as useful to them as they are to the diurnal lizards and the snakes where the tongue transports scent particles. They may therefore be more concerned with tasting the gecko's food to determine palatability.

Contrary to the general rule that most lizards have movable eyelids, those of most geckos are fused and are immovable, remaining as a clear membrane which they clean by licking. However, the members of the subfamily *Eublepharinae* do have eyelids, and are often called eyelid geckos. Pupil shape in contraction varies in the geckos. The diurnal species have round pupils and the nocturnal geckos have vertically slit pupils which contract almost fully, either straight-sided or lobed, when exposed to bright light. Lobed pupils, which provide stenopaic vision, are those that close vertically but leave a few (usually four) small holes in a vertical line. Each one focuses an image on the retina, which in total stimulates the retina more and even allows these geckos to hunt in daylight. The flying gecko (*Pytochozoon kuhli*) probably has the best eyesight of all the lizards. It has a scalloped pupil, which in contraction leaves just a few pinholes open where the scalloped edges do not meet. In most species the color of their iris resembles their body coloration, so that when their pupils contract in bright light the increased area of the iris is less obvious and does not draw the attention of predators.

Like most diurnal lizards the nocturnal geckos also have multicolored oil droplets (or retinal filters) in the photoreceptors of their retinas, and these are known to be beneficial for color vision, as they increase the range of discernible color. So the geckos are assumed to have a degree of color vision, at least when they are active by day, for not even the greatest color vision is of any value at night. While they cannot see a night scene in color any more than we can, some may therefore have a degree of color vision if they open their eyes in daylight. Unlike the snakes in which a filter protects the eye from ultraviolet light, opsins or retinal proteins in the cone receptors of the nocturnal geckos' eyes are calibrated to discern higher wavelengths, into the ultraviolet range beyond human vision. Vision accommodation[5] is achieved by stretching the lens with the ciliary muscles, which relax the zonules or fibers that hold it in place, allowing it to change shape. Lizards, unlike the snakes, have a choroid body[6] that projects into the vitreous humor of the eye and nourishes the cornea.

Lizards lack both outer ears and external ears. Their sense of hearing, at least of airborne vibrations, begins at the tympanic membrane or eardrum, which is visible on the surface of the head, and covers the middle ear cavity. Airborne vibrations are detected by the membrane, while ground vibrations are picked up by the quadrate bone, both of which in turn vibrate the tiny bones of the middle ear that pass the information to the cochlea in the inner ear. The inner ear then transmits it to the brain along the auditory nerves. Geckos' ears are less visible, but can be seen as small holes on the sides of the head, and they have well-developed middle and inner ears and good hearing. In association with their hearing sensitivity they are noisy creatures, with vocal chords unlike other lizards, and with loud voices for their size. They communicate with a range of noises—bird-like chirps, clicking, and even barking—which play an important part in their territorial and breeding behavior. The loud voice of the tokay gecko has awakened and startled many visitors to Southeast Asia in the middle of the night.

Tokay Gecko (*Gekko gecko*)

A very popular pet lizard, the 12-inch-(30 cm)-long tokay gecko is one of the largest geckos, a cylindrical species, slightly flattened on top, with a large head and definite neck. It originates from southern Asia and Indonesia, a forest-dweller that lives on tree trunks and cliffs; but it was introduced into Hawaii, Florida, Belize, and several Caribbean islands in the last two decades of the twentieth century. The tokay gecko is a bad-tempered animal, which threatens with an open-mouthed display and is always ready to bite and can draw blood. It is bluish-silver with reddish or orange-brown and white spots, and its name is derived from its call. Tokay geckos are nocturnal and insectivorous and like most geckos their tails break off easily. Their eyes are large and prominent and have vertical slit pupils, and they have the microscopic filaments on their toe pads which enable them to walk up any smooth and dry surface. Males are slightly larger than females but can also be identified by the swollen base of their tail due to the presence of the hemipenes—the lizard's (and snake's) bi-lobed reproductive organ. The tokay gecko's sense of hearing is said to be within the range of 300–10,000 Hz (see color insert).

Leach's Giant Gecko (*Rhacodactylus leachianus*)

This is the world's largest gecko, with a head and body length of 10 inches (25 cm) and a tail of equal length in juveniles, but shortening in adult life to about 5 inches (12.5 cm) long. It has broad, webbed toes and folds of loose skin on the sides and stomach and varies in color, from the typical grayish-green to brown, with lighter brown spots. It is totally nocturnal and arboreal, living in the tree tops where its diet consists of insects and fruit. The giant gecko is endemic to New Caledonia and neighboring Ile des Pins in the South Pacific Ocean, where it is a highly threatened species, due to alien pests such as the rats and cats that eat its eggs and juveniles, and the introduced fire ant, which is now also attacking its eggs. Loss of forest habitat and excessive collecting for the pet trade have also reduced its numbers. Like all the geckos it has a very slow replacement rate, laying only two eggs at a time, which take about eight weeks to hatch.

Leopard Gecko (*Eublepharis macularis*)

A terrestrial lizard of the sand and rock deserts of southern Asia—Afghanistan, Pakistan, and western India—the leopard gecko escapes the intense summer heat by being nocturnal and crepuscular, and spends the hot daylight hours hidden in a burrow or beneath the rocks. It stores fat in its tail and estivates (the equivalent of winter hibernation) during the hottest time of year. It is yellowish-white covered with small dark spots, and reaches a maximum length of 10 inches (25 cm). Leopard geckos are eaten throughout the Far East as they are believed to cure asthma and other lung conditions. In the west they are very popular as pets as they are easy to maintain and breed, and after two decades of selective breeding they are now available in a number of new pattern and color mutations including hypomelanistic,[7] leucistic,[8] and tangerine. Leopard geckos are members of the sub-family *Eublepharinae*, so they have moveable eyelids, and they also have an obvious outer ear; but they lack the filaments on their toes so they cannot walk up smooth vertical surfaces.

Tuataras

A true "living fossil," lizard-like in appearance but not a true lizard, the tuatara is a unique reptile from New Zealand, which has a primitive body structure and is believed to be an old species that has changed little in over 200 million years. Originally thought to be a lizard, its anatomical differences were sufficient for it to be placed in an order of its own, *Sphenodontidae*. Unlike the lizards it has no visible ear openings, and its teeth are unique, with a single row in the lower jaw that fits into a double row in the upper jaw. These teeth are made of bone and attached to the outer surface of the jawbone, and when they are worn down the tuatara can still "chew" its food with its jawbones. Another significant difference is that the male tuatara lacks a penis, which is present in many reptiles. Tuataras have a scaly but soft skin that ranges in color from olive-green to brown, and they have a crest of spines along their backs and heads. Males grow to 2 feet (60 cm) and weigh just over 1 pound (454 g), and the females are slightly smaller. They have a low metabolic rate and are very long-lived, a lifespan of sixty years being common.

Their variable body temperature has allowed tuataras to survive in a temperate region that does not experience very hot summers. They prefer a temperature range of 60–70°F (15.6–21.1°C), and temperatures over 80°F (26.7°C) have proved fatal, so they avoid extremes. Their nocturnal habits protect them as they spend the daylight hours in burrows (although they may appear to bask briefly), and then they avoid low winter temperatures by hibernating. Sedentary by nature, they rarely move far from their burrow entrance when they appear at dusk, when they feed mainly on the eggs and young of the shearwaters whose burrows they occupy. Also, the rich soil produced by the shearwaters' droppings and resultant vegetation attracts and supports a large population of invertebrates, especially the large-bodied, grasshopper-like weta, which is a tuatara favorite.

Like many diurnal lizards, and yet unlike the nocturnal geckos, the tuatara has a "third eye" or parietal eye, which is a photosensitive organ connected by nerves to the pineal body—a small endocrine gland in the brain. Its functions include acting as a light sensor, probably controlling the lizard's circadian rhythms, and it is active in thermoregulation or temperature control and in triggering hormonal production, especially that which is concerned with reproduction. The eye is situated on the top of the tuatara's head and is quite visible in young individuals up to the age of five months, after which it is covered with skin. It does not provide vision in the normal sense, as it has only a rudimentary retina and lens, but it is sensitive to changes in light. Tuataras have a thick, un-forked tongue, unlike most lizards and the snakes, and although earless they are sensitive to vibrations in the range of 100–800 Hz. There are only two species of tuatara, both of which are considered endangered and have been protected for many years.

Northern Tuatara (*Sphenodon punctatus*)

The most plentiful of the two species of tuatara, the northern tuatara lives on islands in the Marlborough Sound, especially Stephen Island, and on islands in Hauraki Gulf, off the Coromandel Peninsula and in the Bay of Plenty. It is olive-brown with pale underparts.

Brothers Island Tuatara (*Sphenodon guntheri*)

This is a very rare lizard, with a population of only about 300 adults on Brothers Island in the Marlborough Sound. It is olive-green with greenish-yellow spots.

Heloderms

Only the Gila monster and the very similar beaded lizard are venomous, with the ability to inject venom, although certain iguana and monitor lizards do have venom-producing organs. They inject venom not in the manner of the snakes with hollow front fangs, but like the rear-fanged species such as the mangrove snake and the boomslang, with grooved teeth that channel venom from the ducts to the bite wounds. Although their bite is extremely painful, the neurotoxic venom from their paired and modified salivary glands is rarely fatal to healthy humans. These lizards are known as heloderms after their genus *Heloderma*, and are the only survivors of a much larger group of venomous creatures that evolved in the southwestern deserts of North America about 40 million years ago. They are stocky lizards, with rounded heads, plump tails, and short but powerful legs, and their skin is granular or finely beaded.

The heloderms are solitary lizards, both crepuscular and nocturnal in their daily activities, that live in desert and semiarid regions, usually where there is scrub cover. They eat invertebrates, lizards, snakes, young rodents, birds' eggs, and nestlings, and can climb into the lower branches of trees and shrubs to raid nests. They can eat a large amount at a sitting—up to one-third of their body weight— and can soon store sufficient fat for their long combined estivation and hibernation, which may last ten months out of each year. They live a mostly subterranean life, appearing on the surface at night in April and May and then retiring to survive on the fat stored in their tails. With such a short active life and low metabolic rate they are long-lived and have reached the age of twenty years. Mating in the spring, they lay up to twelve eggs in midsummer which they bury several inches deep in the sand. The incubation period is five months, but the eggs sometimes overwinter and hatch the following spring. Like the amphibians, birds, monotreme mammals, and other reptiles, lizards have a cloaca—the single chamber into which the digestive, urinary, and reproductive systems empty. In the heloderms, like other lizards of hot environments, when wastes from the kidneys enter the cloaca, most of the water is reabsorbed and semisolid wastes are then excreted via the anus, thus conserving body fluids.

The heloderms have a forked tongue like many of the diurnal lizards (and snakes), and pick up scent particles which are transferred to the vomeronasal

Gila Monster   A native of the southwestern United States and neighboring Mexico, the Gila monster is one of only two venomous lizards in the world. Highly nocturnal, it lives a mostly subterranean life, being active in the cooler nights of spring, and per- haps again in the fall, but otherwise hibernating and estivating underground when it survives on its stored fat.

Photo: Courtesy Indianapolis Zoo

sensory organs and thence to the brain for analysis, so they no doubt rely more upon these chemical sensing powers than on sight when hunting at night. When they hunt in daylight, which they do occasionally, they can also employ vision and their keen sense of hearing.

Gila Monster (*Heloderma suspectum*)

This is the species of the southwestern United States and neighboring north-western Mexico, specifically the Mohave, Chihuahua, Sonora, and Colorado deserts. It is mottled or banded with black, pink, and orange, and its tail has five bands. It reaches a length of 2 feet (60 cm) and may weigh 5 pounds (2.2 kg). There are two races of this lizard: the southern one is the reticulate Gila monster (*H. s. suspectum*) of the Sonora and Chihuahua deserts, in which the adults are more mottled and blotched, with a black face and a background color dominated by black. The northern race, the banded Gila monster (*H. s. cinctum*), lives in the Mohave Desert and is distinctly black-banded on a light-pink background, but also has a black face.

Beaded Lizard (*Helodera horridum*)

The beaded lizard lives in the deserts and canyon country of western Mexico, especially in the arid regions along the Pacific Coast. It is larger and darker than the Gila monster, quite dark in fact, with some specimens being almost completely black, but generally with bands of yellow spots around the body. It grows to a length of 3 feet (92 cm) and reaches a weight of 6 pounds (2.7 kg).

Notes

1. There are also numerous species of lizards that are legless and that superficially resemble snakes, except for their eyelids and ear openings.
2. Scale-less snakes have appeared occasionally in several species as a result of a genetic defect. They have a coarse, rubbery appearance.
3. In North America the wood frog is the only cold-blooded animal living within the Arctic Circle.
4. When two separate images of an object are combined by the brain to create stereoscopic or 3D vision, essential for the perception of distance and depth.
5. The reflex adjustment action of the eye. Focusing on a near subject and then a distant one, and vice versa.
6. An extension of the vascular coat or choroid of the eye between the sclera and the retina.
7. In which there is decreased black pigment.
8. White color variations.

4 Birds of the Night

Compared to the large number of diurnal birds there are relatively few truly nocturnal species, and in only a few families are the members all or mostly active at night. They are the owls, kiwis, stone curlews, the cavebirds, and the nightbirds. Several other bird families, such as the tubenoses, auks, herons, penguins, waterfowl, and even the parrot family contain a few nocturnal species, and some of the most diurnal birds adopt nocturnal habits during their migration.

The nocturnal birds vary considerably in their lifestyle. Most have the reversed activity patterns of a typical night shift, sleeping during the day and actively searching for food and mates at night. Others have a most unusual variation of the normal diurnal bird activity, feeding out at sea during the day, but returning after nightfall to undertake all their nesting duties in the dark (including digging the nest burrow), and then departing to sea again before sunrise. One nocturnal bird rarely, if ever, sees daylight, as it spends the day in the darkness of a deep cave and only leaves it at night to feed.

To cope in the dark the nocturnal birds obviously have highly developed senses, and their sensory systems outperform those of the diurnal birds in all except distance and color vision. They naturally have the best night vision; the kiwis have the best sense of smell of all birds, while the owls have the best hearing in the bird world. Several also possess a sixth sense, called echolocation, for navigating in complete darkness.

Birds generally are highly visual creatures, and sight is considered the diurnal bird's most important sense. In fact, the class *Aves* is the only group of vertebrates that does not have any permanently blind members, as do the mammals, reptiles, amphibians, and fish. Vision is also an important sense for nocturnal birds, but however good their sight, the eye that can see in complete darkness has not evolved, so navigating at such times must involve another sense. Barn owls, for example, can hunt in total darkness only by virtue of their exceptional hearing, and the oil birds navigate in pitch-black caves only through echolocation, using sound waves (sonar) and their acute hearing to steer clear of the walls and find their nests and roosting places. The kiwi has such poor vision that it depends more upon other senses, especially smell.

Birds' eyes differ from those of the mammals. Their retinas do not contain the extensive vascularization (development of blood vessesls) of the mammalian retina, and must therefore be nourished differently, which involves a structure called the pecten. This is a densely pigmented, fan-shaped extension of the retina which projects into the vitreous humor, its rich blood supply providing the retina with diffused nutrients and oxygen. Its main function is therefore said to be a trophic one—concerned with nutrition—but it may increase the eye's sensitivity and aid vision due to the uneven shadow it casts on the retina, in which case it also has optical value. Adding weight to this theory is the fact that nocturnal birds have a very small pecten compared to the diurnal species, which have the greatest visual requirements.

Unlike the colorful plumage of many diurnal birds, nocturnal species are mostly clad in brown, black, gray, and white, and particularly in mottled patterns, which helps to conceal them in daylight. Their retinas have many rods which allow them to make use of low light, but do not permit color vision, and bright feathers would be wasted on them, as they see their world in shades of gray. Hearing is a well-developed sense in many nocturnal birds, especially the owls, cavebirds, and kiwis. The avian ear lacks an outer ear or pinna but does have external, middle, and inner ears like the mammals, and although the short canal (their external ear) that leads to the eardrum is covered with feathers these do not obstruct sound transmission. Birds' middle ears have just a single bone—the stapes—and the inner ear has a cochlea where sound is converted to nervous impulses which go to the brain. The cave-dwellers, which have a "sixth sense," an echolocation system similar to the bats, naturally also have well-developed hearing to quickly receive the echos from their calls, or their system would fail. Although few diurnal birds have a sense of smell, it is well developed in many nocturnal species, especially in the kiwis, shearwaters, and the oil bird. They are called macrosmatic birds—having a well-developed olfactory sense.

Taste is not well developed in birds, whether diurnal or nocturnal, and most species have only about fifty taste buds, compared to 500 in cats and 1,700 in dogs, and so are unlikely to discern a great range of flavors. Most of their taste buds are situated at the base of the tongue (because in most species the tip is hardened) and on the floor of the pharynx. Birds lack teeth and do not chew, so their sense of taste does not function until the food is at the back of the mouth.

The parrots are an exception, as their taste buds are on the front of their thick tongues, and they can be seen tasting their food as they hold it in their foot. However, birds' limited sense of taste is still sufficient to recognize and reject distasteful butterflies, and in tests they have not accepted insects that were injected with pesticides. Hummingbirds can certainly distinguish the different types and concentrations of sugar solutions and reject flowers whose nectar has a low sugar content; they prefer those with at least 25 percent sugar.

Some of the Species

Raptors

Generally, the activity patterns of the raptors are quite distinct. The owls hunt at night and the hawks and falcons by day, but there are some diurnal owls and several small hawks that are mostly crepuscular, active at twilight when their prey (mainly bats and moths) are themselves in the air, and especially when the bats are leaving or returning to their caves en masse. However, owls are the only birds of prey in which most species are habitually active after dark, and are the nocturnal counterpart of the diurnal raptors—the hawks, falcons, and eagles. The most noticeable differences between the owls and the daytime hunters are the owls' large heads, short necks, broad facial discs, and large eyes; and as their eyes are fixed in the sockets they must move their heads to shift their focus. The owl's flattened face forms a characteristic disc and some species have distinct tufts of feathers which are misleadingly called "ears" or "horns" and are therefore often incorrectly associated with hearing. Unlike the diurnal birds of prey, of which several eat vegetable matter, all owls are totally carnivorous.[1]

Although owls are always associated with darkness, several species are actually quite diurnal. The snowy owl is forced to hunt in daylight in the Arctic summer when the sun never sets. The little owl is also almost completely diurnal and even has sufficient cones in its retina for color vision. The burrowing owl is out by day and often sunbathes, and Tengmalm's owl is diurnal like the snowy owl in the far north during the long summer days, but is strictly nocturnal in the coniferous forests south of the tundra.

The owls' large tapering eyeballs provide a wide expanse of retina, and although they cannot see in total darkness they have extreme visual sense, and with a massive concentration of rods in their retinas they have the most well-developed night vision of all birds. The rods are very slender, which allows more to be packed into the retina than in other birds (the tawny owl has 56,000 per square mm). Their large lens and deep anterior chamber permits a longer "throw" of the image, which provides greater visual acuity. Large eyes allow greater expansion of the pupil, and in dim light the owl's iris dilates almost twice as wide as that of a diurnal hawk, allowing more light to enter the eye and producing a brighter retinal image in poor light. As a result a nocturnal owl's eyes are at least ten times more sensitive than the human eye.

Even with all the rods in their retinas, the nocturnal owls also have room for some cones, and therefore have surprisingly good daylight vision. With their wide range of adjustment the owl's pupils prevent excess light from striking the retina in bright daylight, and a great horned owl discovered and mobbed by small birds in broad daylight has no difficulty flying off to find another hiding place. At roost during the day in exposed places, owls can close their third eyelids independently to protect their eyes from sun glare, and they can blink.

Owls have huge eyes, the largest of all animals in proportion to their size, and the eagle owl's eyes are larger those of an adult human. They are so large they almost touch inside the owl's head, and are separated only by a thin septum or partition; they take up so much space there is no room for muscles, so the owls have lost eye mobility for greater visual acuity. Lacking musculature they are secured by bony structures called schlerotic rings. Their eyes face forward and have a visual field of 110 degrees of which only 70 degrees is binocular vision,[2] so they compensate by being able to move their heads through 270 degrees, and can look behind without moving their body. Most birds cannot move their eyes and rely upon the movement of their heads and necks to extend their field of vision, but no other bird has the owl's range of head movement. Their small bills point downward and therefore do not obstruct the visual field. The owls' brains have also suffered from the lack of space, and their eyes outweigh their brains, but birds generally have large eyes that take up so much space in their small heads that there is little room left for the brain—hence the term "bird brain."

Owls lack a fovea (the most sensitive area of the retina), and with their greatly enlarged pupil for night vision have poor accommodation. They cannot focus on both near and distant objects as they cannot adjust the radius of curvature of the lens, so they are shortsighted, which explains why they always hunt close to the ground.

Despite their good eyesight, when owls hunt on a moonless night they rely more upon their highly developed sense of hearing than vision to locate their prey, and they can hunt in complete darkness by virtue of their hearing. They have the most sensitive hearing of all birds, ranging up to 20,000 Hz, and it is very acute at some frequencies, allowing them to hear the tiniest rustling of a mouse in the grass. The owl's acute hearing is a result of having large inner ears, more nerve cells in the auditory region of the brain than other birds, a wide outer-ear tube, and in some species an erectable concha or external ear. The external ear or auditory meatus in most birds is a short, feather-covered tube, but the owl's ear cavity is so large it may extend almost across the side of the bird's head. The erectable concha, a flap of skin that protects the ear and funnels sound waves to it, lies over the ear openings and is covered by feathers called auriculars, which are loose and airy because they lack the barbules that zip feathers together to make them wind resistant. The owl's facial discs funnel sounds to the ears, and have been likened to satellite dishes.

Owls can also pinpoint sounds with great accuracy. The standard vertebrate ear arrangement is in the same position on each side of the head, so that hearing is said to be horizontally symmetrical. This is good for pinpointing sounds originating from a horizontal plane, but cannot quickly locate sounds in the vertical plane. This would be inadequate for owls trying to quickly and accurately locate prey in almost complete darkness, and evolution's answer was asymmetrical ears. An owl's ears differ in size and position on each side of its head, the degree of asymmetry varying among the species according to their activity patterns and their use of hearing as the dominant sense. The barn owl's right ear opening is lower than the left side, so sound from below reaches it first. As the area of sensitivity differs on each side of the head it allows the bird to locate a mouse in both the vertical and horizontal planes, from a single sound. Direction is determined through the minute difference in time (which is measured in millionths of a second) that a sound is received by first one ear and then the other, these signals being combined instantly in the owl's brain to pinpoint its prey.

Barn Owl *The barn owl has the most acute hearing of the owls and probably of all birds. Its large facial disc funnels sounds to its outer ear cavities, which differ in size and position on each side of the head, allowing it to pinpoint sounds with greater accuracy, and to locate a mouse in complete darkness.*
Photo: Chris Moncrieff, Dreamstime.com

The owls' highly developed sense of hearing is not compromised by the rustling or flapping of their feathers in flight. Unlike the heavy and noisy whirring flight of many birds, owls fly softly and noiselessly, due to their small wing-loading—with a large wing area relative to their body weight—and to the structure of their feathers. Their flight feathers have a comb-like fringe on their outer edge and a soft fringe on their trailing edge, which reduce the noise made by air flowing over the wing, and the upper surface of their wing feathers is softened by a downy covering. These enhanced senses make owls the perfect nighttime hunters, but however successful they may be at night, their survival is equally dependent upon concealment during the day, for they have many predators, although generally the larger species are relatively safe. Large owls kill smaller owls and in the tropics the giant nocturnal constricting snakes catch owls. Young owls, which habitually leave the nest before they can fly, are vulnerable to hawks and eagles and to terrestrial mammalian predators when they rest on or near the ground.

Barn Owl (*Tyto alba*)

The barn owl is a cosmopolitan bird and over its wide range has many races. It occurs in the Americas, Africa, Asia, Australia, and Indonesia, and is mainly a tropical and subtropical species that has extended its range into temperate regions, but not into the higher latitudes. It is probably the most recognizable owl, with its distinctive pale coloration of golden-brown upperparts and white underparts, and its large, heart-shaped facial disc, which collects and concentrates sounds onto the ear openings that are hidden beneath the feathers. The barn owl is a medium-sized owl that preys mostly on small mammals—voles, mice, shrews, and bats; plus frogs, small lizards, and birds up to the size of moorhens and jackdaws. It has suffered greatly from loss of habitat and has been the object of several captive-breeding and release programs. It has an eerie, high-pitched screech and is believed to have the most acute hearing of all owls, and therefore of all birds.

Eurasian Eagle Owl (*Bubo bubo*)

Eagle owls are large and powerful birds that are distributed throughout the world with a single representative in North America. Adult birds are 28 inches (71 cm) tall and weigh 6 pounds (2.7 kg). They live only in extensive undisturbed areas of forest, mountains, and open grassland. The largest species is the Eurasian or great eagle owl, which has many races in Eurasia, North Africa, and the Middle East. It is brownish-black and tawny-buff with streaked underparts and a white throat, and has very large orange eyes and prominent ear tufts. It is a solitary bird that breeds very early in the year, even in the north of its range, commencing in January or February, but breeding is based upon food supplies and it does not reproduce if food is short. The female incubates the eggs and the male provides the food while she is nestbound, and both aggressively defend the nest and territory. Eagle owls are very long-lived, and lifespans of fifty years have been recorded, largely due to the fact that they have no natural enemies since they top the food chain in their particular niche. They eat rodents, rabbits and hares, deer fawns, and birds the size of crows and ducks, and are considered beneficial rodent controls. They do not build their own nests, but usurp those of others, especially crows and herons.

Great Horned Owl (*Bubo virginianus*)

The great horned owl is the only eagle owl of the New World, but it has a wide range encompassing most of the land mass from the Arctic to the tip of South America. It prefers woodland, but is also found in farmland and rocky arid canyons. Sooty-brown mottled with grayish-brown above, it has darker cross-barred underparts and a contrasting white collar, but it is geographically variable, ranging from very dark birds in the northern forests to very pale ones in the arid regions. It also has prominent ear tufts and an orange-gray facial disc, and is slightly smaller than the Eurasian species, weighing up to 4 pounds (1.8 kg) and standing 24 inches (61 cm) tall; like most owls the female is slightly larger. It is a very powerful bird, able to kill pelicans and swans, but its diet consists normally of rabbits, squirrels, skunks, rats, ducks, grouse, and quail, and occasionally domestic cats and poultry. It is a very aggressive bird, especially in defense of its nest, which is usually the abandoned or usurped nest of a crow or hawk. It nests early in the year, and the female incubates while the male stands guard and brings her food.

Pygmy Owl (*Glaucidium passerinum*)

A small, stocky owl comparable in size to a waxwing, the male pygmy owl is only 7 inches (18 cm) in length, while the female is slightly larger. It has a speckled breast, white spots and bars over its brown upperparts; its head is finely speckled with white and it has densely feathered feet. The pygmy owl is a northern bird, ranging across Europe and a broad band of central Asia from the Baltic Sea to the Pacific Ocean, where it is common in coniferous forests. It has an undulating, woodpecker-like flight, and does not automatically migrate south for the winter; it does so only if the weather is especially severe. A natural tree cavity or a woodpecker hole are the preferred nest sites, and uncharacteristically for an owl the pygmy owl lays her whole clutch of eggs (numbering up to eight) before commencing their incubation, which lasts twenty-nine days. Also, unlike the young of most owls, the chicks are able to fly when they leave the nest at the age of thirty days. The female incubates the eggs while her mate provides the food, which consists of birds such as songbirds, woodpeckers, and thrushes, plus voles, mice, bats, and insects.

Bat Hawk (*Macheiramphus alcinus*)

The bat hawk is actually a kite—a medium-sized hawk similar to the falcons with a slender body and long, pointed wings. It is brownish-black with a white throat and chest, with bright-yellow eyes, and has a wide range that includes sub-Saharan Africa, Southeast Asia, and Indonesia. It is a very specialized bird of prey—with large eyes and good eyesight in keeping with its activity in low light, and has long, slender legs and very sharp claws for snatching its prey out of the air.

The bat hawk has two short periods of activity at sunrise and sunset when the bats and cave swiftlets, which are its major prey, are highly concentrated as they leave and return to their roosting caves. It actually lives just inside the entrances of the caves, where it perches on ledges. Although the bat hawk's beak is small it has a huge gape and can swallow small bats and birds whole, which it does in flight, so that its hunting is not disrupted. Some years ago a bounty was placed on its head in Indonesia, not to protect the cave swiftlets for their own sake, but in the financial interests of the nest collectors.

Bat Falcon (*Falco rufigularis*)

This is a small, long-tailed sooty-black raptor, just 10 inches (25 cm) long, with a buff-trimmed white throat, a barred black-and-white chest and chestnut belly and thighs. It is mainly a forest bird, a native of Central and South America, which has lost much of its original habitat due to forest clearance in many regions of Amazonia. It hunts bats at dusk as they leave their roosts, and then again at dawn as they return—a small window of opportunity but a very rich one—and is also active at night when it preys upon large moths.

Kiwis

New Zealand contains a greater variety of nocturnal birds, relative to its total species, than any other country, and the most unusual of these is the kiwi—the most unbird-like bird alive, and the country's icon. It lays the largest egg of all birds in relation to its size; it has the best sense of smell in the bird kingdom; it has hair-like feathers and a body temperature of 100.4° F (38° C), slightly lower than most birds. In outward appearance the kiwi has an unusual, plump pear-shaped body, and has been flightless for so long that its wings have degenerated to tiny claws. Its bones are also very heavy, unlike the birds that fly or those that have more recently lost their flight.

The kiwi has small eyes and poor sight, a good sense of touch and hearing, but its sense of smell is highly developed. It has a large olfactory bulb at the base of its long and narrow bill, and unlike all other species its nostrils are situated at the end of the beak—at the top of the upper mandible near its tip—which is an adaptation for sniffing out worms in the soil and leaf litter. It sneezes and blows a lot to keep them clear of debris when probing the soil, and when checking for airborne scents it holds its bill straight up into the air and sniffs. Its powers of smell are so well developed it can walk with its bill just skimming the surface, prodding down into the soil when a worm is located. The tips of their bills are also very sensitive and they have tactile facial bristles. The kiwis' keen sense of hearing no doubt also aids them in locating worms, and is certainly involved in communication; the males have a high-pitched call and the females have a guttural tone. Their nocturnal habits possibly developed as a result of food availability, as worms migrate closer to the surface at night, and no doubt to avoid New Zealand's now extinct large raptors. Their concealment during the day in deep burrows has provided some protection from the many introduced predators such as ferrets, stoats and brush-tailed possums, and feral dogs and cats, but their populations have been severely reduced, and predator-free offshore islands now provide the only secure habitat for their dwindling numbers.

North Island Brown or Mantell's Kiwi (*Apteryx mantelli*)

This kiwi shows a great range of color variation, and albinism frequently occurs. Generally, its head and neck are blackish-gray and its upperparts are dark rufus streaked with black, but in some birds the black is predominant, while in others there is so little black streaking that the rufus feathers are more prominent. The underparts are pale grayish-brown, the bill is light horn color, and the feet are brown. It survives in the major forested regions and adjacent scrub areas in the northern half of North Island, where locally it is still a common bird. It occurs also on Little Barrier Island, which had an original population prior to others being translocated there. The saving of many of New Zealand's endangered species has depended upon the availability of islands free of predators, and North Island brown kiwis have also been introduced onto Kawau and Ponui Islands in Hauraki Gulf, Motorua Island in the Bay of Islands, and Kapiti Island, northwest of Wellington, where the earlier introduced possums have been exterminated. Ten North Island brown kiwis released on predator-free Motukawanui Island in Matauri Bay in 1995 had increased to fifty birds by 2004. They were part of the program called Operation Nest Egg, which involved the removal of eggs and chicks from the nests of wild birds, and raising them in captivity until they were capable of fending for themselves.

Great Spotted or Large Gray Kiwi (*Apteryx haasti*)

This is the largest kiwi, in which hens weigh up to 6 pounds (2.7 kg) and are almost double the size of the males. The great spotted kiwi's head and neck are dark grayish-brown, and its upperparts are fulvus, mottled, and banded with brownish-black, giving it a spotted appearance. Its gray underparts are also similarly mottled. It is a native of South Island, on the west side of the Southern Alps, where it is split into three distinct populations. The largest of these lives in northwest Nelson, especially in Kahurangi National Park, with others on the northern west coast, particularly in the Paparoa Range, and a third group in the Southern Alps from Lake Sumner to Arthur's Pass. It is a very hardy bird, which is now most plentiful in wet and mossy subalpine areas, for in lowland and coastal forest it has been preyed upon heavily by dogs, pigs, and stoats and has declined in many places; however, these predators rarely venture into the higher altitudes. As its present distribution is mostly in protected areas it is not threatened by habitat destruction. It is believed to have always been restricted to South Island, despite the land bridge

Great Spotted Kiwi   *With poor eyesight, smell is the kiwi's most important sense. Unlike all other birds, its nostrils are at the end of the beak, and it probes in the soil to locate the earthworms that are its main source of food.*
Photo: Orotohanga Kiwi House

between it and North Island, which existed until the end of the last Ice Age. In 2003 the current population of this species was estimated at about 17,000 birds.

Little Spotted or Lesser Kiwi (*Apteryx oweni*)

This kiwi was a common bird in the middle of the nineteenth century, but it is now one of the rarest. It is also the smallest of the kiwis, with the larger females weighing up to $4\frac{3}{4}$ pounds (2.1 kg) and the males just $2\frac{3}{4}$ pounds (1.2 kg). It has yellowish-gray feathers on its head and neck, and its body is similar but lighter and is banded and mottled with blackish-brown, which produces the spotted effect. The bill is flesh-colored and the feet are off-white. Its habitat is woodland and tussock grassland, originally on the western side of South Island from northeastern Marlborough to southern Fiordland and on D'Urville Island. It also occurred on North Island, as a few birds were discovered in 1875 on Mt. Hector near Wellington, just below the snow line in mossy subalpine bush, but this colony has long since vanished. It has also disappeared from D'Urville Island and is now also believed extinct in its original South Island habitat. Fortunately, it was introduced to Kapiti Island early in this century, where the population now numbers almost 1,200 birds, despite heavy initial egg predation by weka rails and rats, which have since been totally eliminated from the island. This species was also more recently introduced to Red Mercury, Hen, and Long Islands, which now each have small but thriving populations. After being absent from North Island for over a century the little spotted kiwi has returned there—to Karori Sanctuary in the city of Wellington—where twenty birds were released in 2000.

Stone Curlews or Thicknees

The stone curlews are also called thicknees because of their thickened tibio-tarsal joints. They are large, ground-dwelling plover-like birds with long yellow legs, large heads and large eyes with a bright-yellow iris, and a strong bill. Their plumage is usually streaked sandy-brown and black with paler underparts, which provides good camouflage when they are resting during the day. They are almost cosmopolitan, occurring in Europe, Asia, Africa, Australia, Central America, and South America, where the double-striped thicknee (*Burhinus bistriatus*) is a common species, but they have very definite habitat preferences. They are birds of the open, drier country such as heaths, downs, grassland, bushveldt, and scrub, plus the sandy and stony banks of rivers, lakes, and estuaries. Stone curlews are nocturnal and crepuscular, and make their presence known in the dark with their long, mournful wailing cries. They prefer to run in a crouched position when alarmed, and then fly low over the ground with deliberate wing-beats. Sight is their

Double-striped Thicknees   *A crow-sized ground-dwelling bird of the open country of central and northern South America, the double-striped thicknee's loud cackling cries are heard after sundown. Its large eyes are an indication of its nocturnal activities. It prefers to run from danger rather than take flight.*
Photo: Clive Roots

prime sense at night, and they have a tapetum that reflects light back through the retina for improved use of low light.

Eurasian Stone Curlew (*Burhinus oedicnemus*)

The stone curlew has a wide distribution across southern Europe and Asia, plus North Africa, and is a rare summer visitor to England from its winter home in Africa. Outside the breeding season it is a social bird that usually congregates in small flocks. It prefers open dry areas, such as arable farmland and bare pasture that afford long-distance views. When alarmed it crouches low and runs and only takes to the wing as a last resort. It is one of the smaller stone curlews, 16 inches (40 cm) in length, a buffy-brown bird streaked with darker brown and with paler streaked underparts. Invertebrates form the bulk of its diet, including worms, snails, slugs, beetles, and grasshoppers, plus field mice, frogs, and small lizards. The stone curlew's nest is just a simple scrape in the stones or sand where two eggs are usually laid. It was once a widespread bird in southern England, in farmland and on the Downs—the rolling chalk-based grasslands—but is now extinct in much of its former habitat, and its breeding population is estimated at only 150 pairs. Its demise is due to nest destruction by the new mehanical farming methods, loss of natural grasslands to agriculture, egg collecting, and shooting.

Bush Stone Curlew (*Burhinus grallarius*)

One of the two Australian stone curlews, this species was once distributed widely across the continent except in the rain forests and very arid regions, preferring short grassland and light woodland, with good visibility. Its numbers are now reduced everywhere and it is gone from most populated regions. A highly nocturnal bird, it is rarely seen during daylight because of its cryptic plumage of pale gray streaked with dark brown. It is seldom seen at night either, but makes its presence known with its eerie, wailing calls. The bush stone curlew has a long neck and long, thin legs with thick knees typical of the stone curlews, and its legs fold backward when it sits. When threatened it rarely flies, preferring to run with its neck outstretched and then stopping after a short distance to determine the danger. The bush stone curlew eats snails, insects such as grasshoppers and beetles, and small frogs and lizards. It nests in open and slightly elevated areas where it has a good view of the surroundings, and lays two eggs in a simple scrape in the soil. The parents share the incubation duties; the chicks are precocial on hatching and immediately accompany their parents away from the nest, usually after the adults have eaten the eggshells for their calcium content.

Cavebirds

Echolocation is used by some animals to determine their position or detect their food, through emitting a stream of high-pitched sounds (generally at ultrasonic frequencies beyond human hearing) and acting upon the returned echos. The time taken for an echo to return determines the distance and location of an obstacle or prey. Echolocation allows flight in complete darkness and is used most successfully by the insectivorous bats, and like them several birds also echolocate to navigate the pitch-black caves in which they nest and roost during the day. The tropical American oil bird and several Asiatic cave swiftlets do this by emitting low-frequency sounds, unlike the bats, which use high-frequency sounds, the delay in their returning echos indicating the distance of obstacles such as the cave walls. These birds have a remarkable sense of hearing, probably the best of all birds, after the owls.

Oil Bird (*Steatornis caripensis*)

Oil birds are the extreme birds of the dark, the bird world's equivalent of the bats, roosting by day in the blackness of caves and being active outside at night, so they never see daylight. They have rich-brown plumage with white spots on the wings, and are about 18 inches (46 cm) long with a wingspan of 42 inches (1.07 m). Although they resemble owls in many ways and are closely related to the nightjars, oil birds are totally different, behaviorally and ecologically, from all other birds. They live only in northern South America and Trinidad, and their most important requirement is caves suitable for nesting and roosting by day. Inside their caves they occasionally nest near the entrance in dim light, but most colonies are completely in the dark. They spend the day there, crouching on a ledge or on their nest, and are entirely dependent upon their echolocation systems for finding their way in and out. They are sociable birds that breed and roost colonially and forage in groups; outside at night they do not use echolocation, but make harsh calls and rely on their vision to navigate and to search for the fruits of oil palms and laurels, which they pluck off with their strong, hooked beaks. They also have well-developed olfactory organs, and it is possible that the sense of smell also helps them locate the aromatic palm fruit. Although the oil birds' echolocation system is similar to that of the insectivorous bats, the clicking noises they emit are at the audible frequency of 7 kHz, much lower than the bat's supersonic range, which varies from 20 to 215 kHz. Their sounds are emitted through the mouth, whereas the bats may use their mouth or nostrils.

Gray-rumped Cave Swiftlet (*Collocallia fuciphaga*)

One of the five species of cave swiftlets, of which three use saliva in their nest-making, this species is a widespread bird in Southeast Asia. It is the only one to make its nest of pure saliva, which is most valuable to the restaurant trade. The "saliva" used for nest-making is a secretion of mucroprotein from their sublingual salivary glands; the nests are edible and are the main ingredient of bird's nest soup, an expensive delicacy in Chinese cooking. They are collected by climbers who use vertical joined bamboo poles to reach the nests. Cave swiftlets have similar habits to the oil birds but are less totally nocturnal, as they may feed by day or night.

These small relatives of the swifts are native to the Indo-Australian region, and are especially common in Borneo where huge, cathedral-like limestone caves shelter thousands of nesting pairs. They navigate in the dark caves and locate their nests, which may be 100 feet (30 m) high on the cave walls or ceiling, among hundreds of others, by echolocation.

Like the oil birds the cave swiftlets do not have the ultrasonic sensitivity of the bats, and make rattling calls composed of rapid sharp clicks, the frequency of which varies from 1.5 to 5.5 kHz, so they are audible to the human ear. The speed of their response to sounds is far greater than man's, however, so they can fly in the deepest and darkest recesses of the limestone caves. These birds actually have two shifts, a very unusual practice. Swiftlets pour out of the caves at dawn as others return after spending most of the night outside. At dusk, the same pattern is followed, with swifts returning after being outside all day, and those that have spent the day in the caves leave to feed.

Nightbirds

When darkness falls around the world and the diurnal insect-eating fly-catchers, bee-eaters, and jacamars find their roosting place for the night, the nocturnal insectivorous birds make their appearance, or at least make their presence known with their repetitious and doleful calls, as they are rarely seen. In the jungles of Venezuela the melancholy notes of the potoo alternate with the roaring of the howler monkeys; in the Australian bush the frogmouth makes a nasal grunting call, while in North America the whip-poor-will repeats its name continuously, a thousand times each night, to the dismay of campers. Nightjars of several species in Asia and Africa make their awakening known with their monotonous calls. All of these birds belong to the order *Caprimulgiformes*, the members of which are mostly nocturnal in their activities. There are four families within the order—the nightjars and nighthawks, the potoos, the frogmouths, and the owlet nightjars; and they are collectively called nightbirds. They have good hearing and excellent night vision, and their far-carrying calls act as a means of communication as they are mainly solitary birds.

Whip-poor-will (*Caprimulgus vociferus*)

Named for its repetitive call, the whip-poor-will is a member the nightjar family, *Caprimulgidae*, which means "goatsucker" in Latin, as in ancient Greece these birds were believed to suck milk from goats' udders, but were of course merely attracted to the goat herds by the insects that accompanied them. Their distribution is central North America from Saskatchewan to Louisiana, where they are rarely seen, being active at night and sleeping by day on the forest floor where their cryptic plumage of dark brown with a black throat matches the dead leaves. Whip-poor-wills are aerial insect hawkers, about the size of a robin, but with a more slender body, long, tapering swept-back wings and a long tail, and they are soft-feathered like the owls. Their bill is small but their mouth is huge, with ample rictal bristles to aid insect capture. Like all the nightbirds their feet are tiny and weak. They fly somewhat erratically at night with heavy wing-beats followed by gliding, wheeling about with sudden twists and turns, as they pursue insects, which they scoop into their large mouths. Whip-poor-wills prefer open country— grassland, heath, semideserts—and in contrast with the upright perching frog-mouths, they rest horizontally along a stout branch or on the ground among the leaves. They do not build nests, and simply lay their two eggs on the ground.

Common Potoo (*Nyctibius griseus*)

Common potoos resemble large, long-tailed nightjars and are the common nighthawks of Central and South America, where they favor open areas such as grasslands and light woodland. Typical of the nighthawks they are not colorful birds, generally being brown mottled with gray, black, and chestnut, with a length of 15 inches (38 cm). Like the frogmouths they sit in an upright position on a tree stump or branch during the day, and elongate themselves even more when alarmed; their cryptic brown-and-black feathering is so successful they appear to be a natural extension of their perch. When "active" at night they sit on a favorite perch and await a passing moth, swooping out like a flycatcher to scoop it, and returning immediately to their perch, just like the diurnal flycatchers and bee-eaters. The potoos have bright-yellow irides that glow orange at night as they reflect artificial light. Like the other nighthawks they have a repetitive call, but it is less mournful, even rather musical on a descending scale, but repeated monotonously. The potoos' nesting habits seem even more precarious than the other nightbirds, for they lay a single egg in a slight depression on a sloping branch and incubate it in an upright position.

Tawny Frogmouth (*Podargas strigoides*)

This bird is the size of a crow, the largest of the fourteen species of frogmouths and the largest nocturnal insectivorous bird. It has a heavy body and a large head with a wide gape similar to a frog's. It lives in Australia and New Guinea in a range of habitat excluding the dense rain forest and deserts. The tawny frogmouth has a grunting call, but also makes a low-frequency sound, which humans may not be able to hear. When asleep during the day it adopts a motionless posture on an exposed branch in a vertical position, with the head pointed upward, and its cryptic plumage of mottled grays and browns blends well with the surrounding branches and foliage and is very difficult to see. Frogmouths become torpid when roosting at night in winter, their body temperature dropping to 83° F (28.4° C) on cold nights, from their normal 104° F (40° C). Although they are nocturnal and normally rest by day, during the winter they search for food briefly just before sunrise and again at sunset and spend the rest of the day and night in a state of torpidity. Their large gape, which was originally believed to be for scooping insects out of the air while in flight, is actually used for plucking their prey off the leaves or from the ground. Their food consists of insects, snails, small mammals, frogs, lizards, and small birds. They have excellent night vision, and their eyes face forward like the owls, so they have good binocular vision. They have large, rounded wings and are swift fliers, and their flight is silent like the owls as they have soft feathers. Frogmouths do not waste a lot of time and energy on nest-building, and the nest is usually just a very flimsy affair of twigs, barely large enough to hold the two eggs.

Tubenoses

The tubenoses belong to the order *Procellariiformes*, which contains a number of marine birds that spend most of their lives at sea. Immature tubenoses rarely come ashore in their first three years, and only begin to make regular annual visits to remote islands when they are sexually mature. They are closely related to the penguins, and unlike all other birds have tubular nostrils—an extension of their nostrils in the form of an open-ended tube on either side of the upper bill. These tubes are used primarily to excrete salt, as the birds must drink salt water during their months at sea, and without this excreting ability the excess salt would cause kidney failure and death.

Most members of the order are nocturnal, but their nighttime activities are mainly restricted to their arrival and departure from their nest sites in the dark, and the activities they undertake there—digging their burrows, egg-laying, incubation, and chick raising. In the family *Procellariidae*, the shearwaters and petrels are nocturnal and the fulmars are diurnal. All the storm petrels, *Hydrobatidae*, and the diving petrels, *Pelecanoididae*, are nocturnal, but the albatrosses of the family *Diomedeidae* are all diurnal.

Tubenoses are one of the few groups of birds with a good sense of smell, with well-developed olfactory lobes connected to their tubular nostrils, which helps them find food on the open ocean. They are expert navigators, converging on their island after months at sea and a journey of several thousand miles.

Short-tailed Shearwater (*Puffinus tenuirostris*)

From its breeding range off the coast of southeast Australia this narrow-winged nocturnal seabird migrates almost 10,000 miles (16,000 km) twice annually, north over the western Pacific Ocean to the Arctic for the northern summer, and then back home down the center of the Pacific Ocean to breed during the Australian summer; a journey that takes about six weeks each way. It nests in vast colonies, with over 20 million birds in almost 300 colonies, especially on the islands of the Furneaux Group, between northeast Tasmania and Victoria. The largest colony, containing 3 million birds, is on Babel Island off the east coast of Flinders Island.

Short-tailed shearwaters dig burrows in which they lay their eggs, and to avoid predators they come ashore after dark and then leave again before daybreak. When navigating toward their tunnels from the sea, the homecoming birds are greeted by the shrill calls of their mates to direct them to the right entrance. Their gregarious nesting habits have not escaped attention, and their fat chicks have been collected from their burrows in large numbers since the arrival of the first colonists, and even before then by visiting sealers, whalers, and shipwrecked or marooned mariners. Even now several hundred thousand of the plump squabs, which are called muttonbirds, never make it to the sea, as they are still collected for human consumption.

Sooty Shearwater (*Puffinus griseus*)

This shearwater is the "muttonbird" of New Zealand, the traditional harvesting of the fat squabs being the exclusive right of the Rakiura Maori. It is still practiced, with 500,000 taken annually from their burrows on the islands off Stewart Island. Shearwater chicks grow rapidly, and are soon double the size of their parents; but before they are fully fledged and able to fly they are abandoned by the adults, which gather at sea in large flocks or "rafts" of several million individuals, during which time they moult their primary feathers and are flightless for several weeks. When they have regrown their flight feathers they migrate north. Their down-covered chicks live off their fat stores for three to four weeks after they have been abandoned and lose weight rapidly, but while doing so they are surprisingly able to grow their own flight feathers, which are made of pure protein. When these are fully grown they also take off for their long migration north to Alaska, where they spend the northern summer with the adult shearwaters.

Auks

The auks or alcids (*Alcidae*) are a family of seabirds with compact bodies, large heads, short wings and tail, and a stout bill. They include the puffins and razorbills, the guillemots and murrelets. They live in the seas of the Northern Hemisphere, and have been called the penguins of the north, but they are not flightless, although flight seems a great effort for their small, narrow wings, which they beat with a whir. Some members of the family undergo a complete wing-feather moult at sea and are flightless for several weeks. They actually spend their whole lives at sea except for the short period on land when they are nesting.

Most auks are diurnal in their activities, nesting high on cliffs in crowded colonies, and laying their eggs directly on the rock. Others nest colonially in burrows, and several of these are nocturnal, resembling the shearwaters in the manner of their arrival and departure from the nest sites in the dark. The most unusual nocturnal nesting activity is shown by the marbled murrelet, which lays its single egg on the branch of a fir tree miles from the sea.

Rhinoceros Auklet (*Cerorhinca monocerata*)

This is a grayish-black seabird about 15 inches (38 cm) long with a white belly, and during the breeding season two white stripes on the side of the head, one above and one below the eye, plus a short, erect horn at the base of the bill. The rhinoceros auklet is a bird of the northern Pacific Ocean, which breeds along the coast from California to Alaska and then down the coast of Asia to Japan and Korea. Although still a common bird in Alaska, it was extirpated in California in 1860 but has since repopulated some of its former range. It is rarely seen because it stays far out at sea all year except for the breeding season, when it comes ashore at night to compete for a nest site, to dig the nest burrow with its bill and feet, lay and then incubate its single egg, and raise the chick. Rhinoceros auklets eat small fish like sand lance, smelt, and young herrings, which they catch after an underwater chase, using their wings for propulsion.

Ancient Murrelet (*Synthliboramphus antiquus*)

This bird with the most unusual name lives in the waters of the northern Pacific Ocean from China to British Columbia. It is a small gray-and-white seabird, about 10 inches (25 cm) in length, with a black cap, a white stripe over the eye, and a yellow bill. Ancient murrelets nest in rock crevices, under shrubs and in burrows which they dig themselves and which may be several feet long; more than half the total population of the species nests on the Queen Charlotte Islands. They are entirely nocturnal on land, arriving and departing in the dark and remaining concealed on their nests during the day. The parents share the incubation of the two eggs, the relieving partner arriving in the dark to take over, and while the adults are nesting the unattached young males perch high in the Sitka spruce trees all night, singing to attract a mate.

The parent murrelets have solved the problem of bringing food, mainly zooplankton, to their chicks in the nest. Two days after the chicks have hatched the parents call them away from the nest in the dark and lead them through the ferns and undergrowth and over fallen tree trunks down to the shore and then far out to sea. The chicks have waterproof down and a layer of insulating fat, and they grow and fledge at sea.

Parrots

Only two members of the parrot family are nocturnal, and one of these, the night parrot, is a very rare bird which has not been seen for many years and may possibly be extinct.

Kakapo (*Strigops habroptilus*)

The kakapo is the world's largest and most unusual parrot. It weighs 8 pounds (3.6 kg), is 25 inches (63 cm) long, and has several characteristics of the owls—including a facial disc of bristle-like feathers, soft plumage, and nocturnal habits. It has bright-green upperparts and yellowish underparts, all barred and streaked with brown and yellow, and a yellow stripe over its eye. The kakapo is a woodland bird, which originally occurred from sea level to 4,000 feet (1,220 m) in the mossy beech forests, especially where they adjoined snow tussock meadows or river flats, of all three of New Zealand's major islands. It was once very common, but with human persecution, introduced predators, and loss of habitat was believed extinct by 1949. A few birds were later discovered in Fiordland and on Stewart Island, and through the efforts of the Department of Conservation and the Kakapo Recovery Program, which has involved the translocation of all the kakapos to predator-free islands, in August 2004 the kakapo's population numbered eighty-three.

During daylight the kakapo shelters in burrows, crevices, among tree roots, or in dense low vegetation. It is almost totally vegetarian and eats a wide range of native and introduced vegetation including shoots, leaves, nuts, berries, fern roots, fungi, moss, and insect larvae. Despite its large, strong wings it can no longer fly because its flight muscles have completely degenerated, and its breast keel is now just a low ridge. A well-developed voice and a good sense of hearing are generally complementary, and as the kakapos make deep booming calls and discordant shrieks and croaks, they are believed to have good hearing. Despite its owl-like appearance, the kakapo has small eyes and poor vision, and its stiff rictal bristles aid its nocturnal search for suitable low-growing vegetation. It also has a taste for nectar and is believed to have an adequate sense of smell to locate flowers.

Penguins

Penguins are usually associated with the snow, icy water, and pack ice of the Antarctic continent, but most species actually live in slightly warmer waters farther north; others reach the tropics and never experience harsh conditions. The other misconception about penguins is that they must be totally diurnal, as photos and film nearly always depict them on bright sunny days. However, like the shear-waters and petrels, which spend long periods and in some cases most of the year at sea, several penguins are also nocturnal in their land activities. They arrive at their nest sites after nightfall, conduct all their breeding activities—preparing the nests, laying, incubating their eggs, and then raising the chicks—under cover of darkness, and then return to the sea before daybreak.

Blue, Little, or Fairy Penguin (*Eudyptula minor*)

The blue penguin is a warm-latitude species that lives on the coasts of southern Australia, Tasmania, and New Zealand. It is the smallest of the penguins, weighing just over 2 pounds (920 g) and is 16 inches (40 cm) long. It has typical white penguin underparts, but its back and head are the palest of all species, being metallic slate-blue, which takes on a brownish hue as the bird ages. Three sub-species are recognized. The southern race (*Eudyptula minor novaehollandia*) nests on the coasts of southern and southwestern Australia and Tasmania and is a deeper blue than the northern race (*Eudyptula m. minor*), which breeds on the coasts of

New Zealand's three main islands and also on Chatham Island. The third race is the white-flippered penguin (*Eudyptula m. albosignata*), sometimes considered a separate species, which breeds only on the Banks Peninsula near Christchurch, on Motunau Island, and on the coast of north Canterbury. It is now a rare bird with a current population of less than 4,000 pairs.

Blue penguins are nonmigratory; they fish in summer in the seas close to their nest sites and return to their burrows to relieve their mates or feed the chicks each night under cover of darkness. They are therefore rarely seen except where they have become a tourist spectacle, such as at Tasmania's Bruny Island and the famous "penguin parade" colony on Phillip Island near Melbourne, where the birds return to their nests under floodlights, watched by people on bleachers. Blue penguins generally nest in sand burrows, which may be 10 feet (3 m) long, sometimes close to the beach, while others trek up to a mile inland, often crossing roads to get home. They are also opportunists and nest in caves, crevices, under dense vegetation, beneath buildings, and under wharves, even in Hobart's busy harbor. Nest boxes have been provided for them at many colonies to increase their chances of breeding success. During the winter, nonbreeding season, blue penguins go farther out to sea to fish and may be away for several days.

Chinstrap Penguin (*Pygoscelis antarctica*)

The chinstrap penguin is another species that has nocturnal habits, feeding at night as well as during the day. It has a bluish-black head with a white face and throat and a thin curved line of black feathers under the chin. It is one of the most abundant penguins in Antarctic and sub-Antarctic waters, with an estimated population of about 7.5 million breeding pairs, and enormous rookeries on the shores of the Antarctic Peninsula and on islands in the Scotia Sea—South Georgia, South Sandwich, and South Shetland. The largest colony, on Zavodovski Island in the South Sandwich group, consists of 10 million birds, and there are smaller rookeries on Heard, Kerguelen, and Bouvet Islands. It has "wandered" at sea as far as Tasmania and MacQuarie Island. They often live on icebergs from which they dive to catch fish and krill, but they return to land to breed on rocky shores, where they make a "nest" of pebbles. Their chicks leave the nest when they are feathered at twenty-eight days old and join together in creches, where they huddle for warmth and protection while their parents are away at sea.

Herons

The heron family *Ardeidae* includes many waterbirds, which are usually divided into four subfamilies. These are the nocturnal night herons, their counterparts the day herons, the tiger herons, and the bitterns. In both latter groups there are nocturnal species, but generally the behavior of these birds is not as definite as their taxonomic divisions. The black-crowned night heron is strictly nocturnal, yet the yellow-crowned night heron (*Nyctanassa violacea*) is actually quite diurnal, and two common species of bitterns—the American bittern (*Botaurus lentiginosus*) and the least bittern (*Ixobrychus exilis*)—are active as much during the day as they are at night. The following species, however, are quite nocturnal in their habits.

Little Bittern (*Ixobrychus minutus*)

This is the smallest of the bitterns, just 14 inches (35 cm) long, which breeds in Europe and western Asia, and migrates to Africa for the winter. It is a shy and elusive bird of the dense vegetation of marshes, river banks, and reed beds, which flies with fast wing-beats followed by long glides; but it does not flush easily and relies on its stealth and stillness for concealment during the day. It perches on the side of a reed stem and stretches its body upward, parallel to the stem. A strongly nocturnal bird, it remains hidden in the reed beds during daylight, and hunts for insects, frogs, fish, and crustaceans after dark. The little bittern has a black back and crown, and a buffy-white face and underparts. Its neck is short and it has a long, pointed bill. Although bitterns generally have booming calls that can be heard up to 3 miles (4.8 km) away, this species makes a deep croaking call. It nests near or over water, making a platform of reeds on which it lays up to eight eggs.

Black-crowned Night Heron (*Nycticorax nycticorax*)

A stocky, short-necked heron, with a black crown and back, gray body and whitish underparts, the black-crowned night heron in breeding plumage has two long, white plumes on the back of the head. Immature birds are dull grayish-brown with white spots. It is thoroughly nocturnal, and begins its search for food at dusk and continues throughout the night. It has large eyes and exceptional night vision, standing motionless in the dark waiting for fish, frogs, and insects to come within reach. It also stalks in grassland, searching for field mice and voles, and raids nesting colonies of other herons, ibis, and terns, whose chicks stand little chance against its stout, pointed bill. The night heron is a bird of both freshwater and saltwater marshes, mangrove swamps and reed beds, and is virtually cosmopolitan in its distribution. It occurs on all the continents except Australia. It is a colonial nester, occasionally in colonies with other herons.

Boat-billed Heron (*Cochlearius cochlearius*)

This is a very nocturnal relative of the night herons that frequents mangrove swamps, freshwater marshes, and river banks from Mexico south to Argentina. The boat-billed heron has a short but very broad bill, and is a medium-sized heron, about 21 inches (53 cm) in length. It has excellent night vision and hearing, and communicates with vocal calls and mechanical sounds made by "popping" its bill. It

Black-crowned Night Heron *With large eyes and exceptional night vision the black-crowned night heron is one of the most nocturnal waterbirds. It stands motionless in shallow water waiting for fish and frogs to come within reach, stalks in grassland searching for mice, and steals other birds' chicks from their nests.*
Photo: Photos.com

has a black crown, crest, and upper back and its wings are pale-gray; it has a white face, throat, and breast, and a rufus belly. When displaying it spreads its crest out like a fan. The "boat-bill" remains concealed in the vegetation during daylight, and is active entirely at night, scooping up shrimp and small fish in its beak. Like all the herons, it has two patches on its breast that produce "powder down feathers," which turn to powder when crushed in the bird's beak. This is applied to soiled feathers and then scraped off with the serrated claw on the middle toe.

Ducks

Few of the 160 species of waterfowl are nocturnal, but several species of ducks do feed at night and rest during the day. A number of these are whistling ducks of the genus *Dendrocygna*, which were formerly called tree ducks, and several members of the genus *Anas*, including the brown teal (*Anas chlorotis*) and the northern pintail (*Anas acuta*). The ruddy duck (*Oxyura jamaicensis*) is also largely nocturnal.

Black-bellied Whistling Duck (*Dendrocygna autumnalis*)

Also known as the red-billed whistling duck, this common species is very active at night, and usually roosts by day in the dead branches of tall trees. A bird of the New World, it ranges from the southern United States through Central and South America and Trinidad, where it lives in coastal regions, freshwater marshes, and swamps. It is a vegetarian, and feeds on the leaves, roots, and seeds of aquatic plants, but is also very destructive in cropland (especially in rice paddies) and is hunted in response. The black-bellied whistling duck lives up to the old name of tree-duck as it does nest in tree cavities, whereas several species do not. It lays up to fourteen eggs which are incubated for twenty-eight days, and the ducklings have to drop down to the water soon after they hatch. This species is distinctively plumaged, with a black abdomen, chestnut wings and back, and a large, white patch on the wing coverts. But the duck's bill is the best identifier, for it is the only whistling duck with a bright-red beak.

New Zealand Brown Teal (*Anas chlorotis*)

A small, brown surface-feeding dabbling duck with chestnut breast, white collar, and white eye-ring, the brown teal is one of the rarest waterfowl. The sexes are alike in plumage, but male brown teal have a high-pitched whistle and females have a more duck-like quack. Once common, like so many of New Zealand's birds, it was reduced to endangered status by a number of threats, including wetland drainage, shooting, alien predators, and the introduction of bird diseases. It has been the subject of an intensive breeding and reintroduction campaign by New Zealand's Department of Conservation, and has been returned to several parts of its former range. Originally a bird of lowland wetlands, it is now mainly a coastal bird, forced into this niche by the loss of habitat and predation inland. It now survives mainly in estuaries and sheltered coastal bays of eastern Northland and several offshore islands, especially Great Barrier Island. There it frequents the shoreline at night, scooping aquatic invertebrates off the surface and probing the seaweed for small crustaceans and molluscs. The brown teal is a poor flier and is the ancestor of two flightless ducks—the Auckland Island teal (*Anas a. aucklandica*) and the Campbell Island teal (*Anas nesioti*), both restricted to sub-Antarctic islands south of New Zealand.

Migrating Birds

Despite having poor night vision, many diurnal birds adopt nocturnal habits when they migrate. On both hemispheres, during the spring and fall there are millions of birds in the night sky, flying along well-recognized routes between their summer and winter habitat. Most songbirds migrate at night, and typical night migrants include sparrows, warblers, robins, wrens and vireos; plus rails, waders, Canada geese, and American bitterns.

One reason suggested for flying at night is the protection factor. Navigating birds usually do so in flocks of others of their kind, and call to each other to maintain contact. To do this in daylight would be an invitation to hawks and falcons to station themselves high above the flight routes, ready to swoop down at will. Consequently, many birds, even strictly diurnal species, migrate under cover of darkness when the high-flying predators are roosting. Another reason to migrate at night may be the nature of the night air. The atmosphere is more stabilized, and the birds have a steadier stream of air in which to fly. The air is also cooler then, and lowered temperatures are an obvious advantage. Flying long distances generates heat and at night there would be less evaporation of their body fluids in the cool air. When flying in warmer daytime air, birds lose heat to maintain their correct body temperature, and they do this through the evaporation of water from their air sacs as they breathe. It has also been suggested that birds flying in the dark may benefit from the celestial navigation—by the stars—but during daylight there are even more aids to navigation such as the sun, the earth's magnetic field, and polarized light.

The most logical reason for night migration is that it gives birds the opportunity to rest and feed during the following day. Even if they stock up with food prior to their flight, and some small birds are known to almost double their weight before setting off, this will soon be depleted because flying long distances uses a lot of energy. The typical nocturnal migrants are all diurnal feeders; they cannot fly all day, rest at night, and fly off again the next morning, as this demands complete reliance on their stored energy for a long journey, with no facility to feed or quench their thirst. So they fly at night and then rest, feed, and drink the following day, and if the landing site is favorable they may even stay there for a few days. Thus, they are able to replenish their fat stores, or at least reduce the total reliance on stored energy for their journey.

Notes

1. It has been reported that the burrowing owl may eat seeds and the fruit of the prickly pear.
2. Humans have a field of view of 180 degrees, of which 140 degrees is binocular.

5 The First Nocturnal Mammals

■ MONOTREMES

The first mammals began evolving from reptiles about 185 million years ago in the Mesozoic Era and initially they were of two kinds, the Therians and the Prototherians. The Therians were very successful and evolved into the present-day live-bearing mammals, now in the subclasses *Metatheria* (marsupials) and *Eutheria* (placentals). The Prototherians were a less successful group and only three survived—the duck-billed platypus and two echidnas—which are included in the order *Monotremata* and are usually called monotremes. They are therefore the most primitive mammals alive today, relics with characteristics of their reptilian ancestors as well as certain mammalian features. Like the higher mammals, the monotremes have a coat of hair, mammary glands for milk production, and they suckle their young. Yet they are the only mammals to lay eggs, which resemble the reptile's, with leathery skins and large yolks. Unlike the mammals—and like the reptiles and birds—they have a cloaca, which is the single opening for passing fluid and solid wastes, for mating, and for laying eggs.

Monotremes resemble reptiles in the structure of their eyes, and in certain aspects of their skulls, ribs, and vertebrae, but their skeleton also has mammalian features, such as a single bone on each side of the lower jaw (whereas the reptiles have several) and three middle-ear bones, unlike the reptile's single one. They also have a typical mammalian four-chambered heart, whereas most reptiles' hearts have only three chambers. Young monotremes have teeth, but they lose them as they mature and the adults are toothless and must crush their food between their hard palates and tongues. To avoid the reptile predators long ago as they were evolving they became nocturnal, as they are to this day.

Duck-billed Platypus

The duck-billed platypus (*Ornithorhynchus anatinus*) is one of a kind, a single species in its own genus, an odd-looking animal, but a very successful one, for fossils show that it looks the same today as it did many millions of years ago. It has the streamlined body of an otter with a short, dense coat composed of a woolly underfur and blade-like guard hairs. It is blackish-brown with pale-yellow or brownish-white underparts; and has short, stout limbs and broad webbed feet, a beaver-like tail, and a leathery snout resembling a duck's bill. It lives in the rivers and ponds of eastern Australia, from Victoria to northern Queensland, and in Tasmania, where it is active throughout the night.

Ornithorhynchus is an excellent swimmer and diver and swims mainly with its forefeet. It cannot stay submerged for long, however, generally just a minute or two when feeding, but up to six minutes if it anchors itself—wedged beneath a submerged log, for example—to stop it bobbing up to the surface. It probes in the river and lake bottoms for aquatic invertebrates, tadpoles, small fish, freshwater crustaceans, and snails, and holds its catch in cheek pouches until it returns to the surface to breathe. Horny plates in each jaw, ridged in front and smooth at the rear, crush the platypus's food, and these plates are continually growing as they are worn down by the grit scooped up with the prey.

Platypuses live in burrows on the banks of rivers and ponds, the diameter of their tunnels being just wide enough to squeeze water from their coats before they reach the nest. They normally keep two burrows active, and their entrances are always above water level. One is for general living for the pair; the other is used solely by the female as a maternity den. She lays two tiny eggs, the size of house sparrow's eggs, in a nest of damp leaves in her burrow. The incubation period is only ten days and she curls around them to provide the correct temperature. The babies are then "brooded" and suckled by her for four months before they appear fully furred at the burrow's entrance. For safety from predators, she plugs the entrance with soil while she is caring for her young.

Unlike echidnas, the platypuses do not hibernate, but remain active all winter even in water at at 32°F (0°C), as the air trapped in their underfur and between the guard hairs provides good insulation. Like the echidna, the platypus has a very low body temperature, several degrees lower than the mammalian average. The male platypus is the only venomous mammal—able to actually inject venom—and has a hollow spur inside each ankle which is connected to a poison gland. The venom can kill a dog but is not known to have caused any human deaths. All baby platypuses have spurs but the females lose them as they mature.

Platypuses lack outer ears or pinnae so hearing is not their best sense. When submerged they cover their ears and eyes with folds of skin anyway, so they are effectively blind and deaf then and rely completely on their sense of touch. Their bills are extremely sensitive organs, which they sweep about to locate their food, and in addition to its tactile sense it is now known that the platypus bill also functions as an electroreceptor, detecting the weak electric fields emanating from its small animal prey, and able to locate them in the mud and under rocks.

The special olfactory receptors known as vomeronasal or Jacobsen's organs, which are present in many reptiles, are well developed in the platypus. They contain sensory neurons similar to those in the nose, which detect chemical compounds, including pheromones that carry messages between animals of the same species and which are connected to the olfactory nerves. The platypus's major predators are carpet pythons, the large monitor lizards called goannas, water rats that kill the babies in the nest, and foxes and feral cats that kill juveniles when they leave the nest burrow.

Echidnas or Spiny Anteaters

Echidnas resemble giant hedgehogs. They have compact, muscular bodies covered with coarse hair and spines, short legs, big feet and large claws, small ears and stubby tails, and their snout is modified to form an elongated beak. They are primarily nocturnal, but they may appear in the early evening, and occasionally earlier on dim days. When threatened they dig into soft soil and curl into a ball, or wedge themselves into a crevice or burrow by erecting their spines. Both males and females have spurs on their hind legs, but they lack the platypus's venom glands. Echidnas cannot regulate their temperatures as well as other mammals, and although they are warm-blooded their body temperature only averages 89.6°F (32°C) compared to the marsupial's 95°F (35°C) and the higher mammal's 98.6°F (37°C). The short-beaked echidna reaches this low body temperature when it is active, and when inactive in cold weather it can drop to 50°F (10°C) for extended periods; it hibernates for several months in the cooler regions.

Echidnas eat mostly termites, ants, and worms, which they dig out with their powerful claws, and then grind between the horny striations on the back of their long, sticky tongues and the hard ridges on the palate. They have small eyes and poor vision and locate their food with their acute sense of smell, and in the case of the short-nosed echidna, at least possibly with the help of electroreceptors in its beak, which are thought to detect the signals given off by its prey. Smell is apparently also important for monitoring their environment and for social encounters with conspecifics.

Echidnas also have good hearing, their external ears being well developed although they are almost hidden by hair, and like many reptiles they are very sensitive to ground vibrations, which they detect with their beaks. These are quite remarkable organs, with receptors for touch, temperature, and electrical impulses; and like the platypus they have well-developed vomeronasal organs for detecting airborne chemical compounds.

Echidnas are solitary animals that socialize only at mating time. The single small egg is retained in the mother's body and provided with nutrients for some time before being laid, when she must maneuver it into the pouch with her beak, the pouch being just a temporary one for the breeding season, which develops just before the egg is laid. It hatches after an incubation period of ten days, and the tiny embryo is naked, blind, and helpless, but is able to suck the very rich milk secreted by glands that open at the base of specialized mammary hairs (the forerunner of mammary glands) onto patches on the mother's belly. When the spines begin to sprout at eight weeks, she leaves her baby in a burrow and returns once a week to feed it, and at this stage of its life it is very vulnerable to predators, especially water rats.

Short-beaked Echidna (*Tachyglossus ecaudatus*)

This species ranges throughout Australia and Tasmania and also occurs in southern New Guinea. It reaches a weight of 13 pounds (5.9 kg) and is 14 inches (35 cm) long, with spines measuring about 2½ inches (6 cm) in length. Its small eyes are situated at the base of the snout and look ahead, providing binocular vision, although not very acute sight. The short-nosed echidna lives in a wide range of habitat, including forest, grassland, and rocky areas, where it feeds mainly upon ants and termites which are gathered with its long and sticky tongue. It suffers heat stress at temperatures above 95°F (35°C), and hides by day in a burrow or rock crevice, appearing late in the afternoon; even at night it avoids high temperatures

Short-beaked Echidna   *With small eyes and poor vision, the echidna relies upon its acute senses of smell and hearing when it searches at night for the ants and termites that are its main foods. It is assisted by the electroreceptors in its highly sensitive snout or "beak."*

Photo: Courtesy www.ausemade.com.au

by returning to its burrow. Its low temperature is controlled by shivering, unlike the nonshivering thermogenesis[1] of higher mammals.

■ MARSUPIALS

The marsupials follow the monotremes as the second most primitive order of mammals, but although they are also called pouched mammals, the pouch or marsupium is absent in the females of some species and in others is rudimentary, being little more than a fold of skin on either side of the nipples. A more appropriate name for these animals is "implacentals," as the common factor linking them all is their very short-lived placenta—through which the higher mammals nourish their embryos in the womb—which contributes little to the marsupial's fetal development, so the premature young are born after a very short gestation period. Their development is then completed "externally" in the pouch or in the absence of a pouch just dangling from a teat, which swells in their mouth to firmly anchor them.

Marsupials occur naturally in only two regions of the world: Australasia, which is their stronghold, and the tropical parts of the New World, with just one species ranging into North America. This discontinuous distribution, with none in the intermediate land masses of Eurasia and Africa, resulted from a chain of events that happened very slowly long ago. While the early Prototherian mammals, which were the first to develop from the reptiles, were evolving into the egg-laying platypus and echidnas and other long-extinct species, the Therians diverged into the marsupials (*Metatheria*) and the higher placental mammals (*Eutheria*), and the evolution and distribution of both is entwined with the breakup of Gondwanaland. Acceptance of the continental drift theory—that all land was once one large mass from which the present-day continents drifted away—makes their evolution from a common ancestor more understandable. Continental drift also explains why the marsupials live mainly on the southern continents,[2] thousands of miles apart, for it is now recognized that the earth's land masses have been considerably redistributed over the geologic epochs.

Marsupial development actually began in North America and they invaded South America while the continents were joined as Gondwanaland. From there they crossed Antarctica (which was a lot warmer in those days) and entered Australia, and were aboard when it broke away and drifted slowly across the Pacific Ocean. This was before the evolution of the higher mammals and especially the carnivores, and their absence gave the primitive, opossum-like marsupials the opportunity to adapt to many diverse niches, a process known as adaptive radiation. Consequently, they vary greatly in form and behavior, and occupy a far greater variety of niches than their ancestors, but the one thing practically all of them have in common is their nocturnal behavior. The marsupials left behind in South America, after its separation from Africa and Australia, did not have such an easy time because they had to contend with the evolving placental mammals, which included cats, dogs, skunks, and raccoons, but several species survived and are little changed from their ancestors. It is significant that they are also practically all nocturnal and are experts in the art of daytime concealment.

Marsupials differ from the higher mammals in a number of ways. Their brains are smaller, but although they have always been considered rather stupid animals, laboratory experiments have shown they are as intelligent as many higher mammals. They have marsupium bones (even the pouch-less ones), which project from the pelvic bone and strengthen the wall of the abdomen to support the weight of a pouch and its contents.

The sense of smell is well developed in the marsupials, important for communication and for food acquisition. They mark their territories with scent from their skin glands, and with urine and feces. Like several other vertebrates, they actually have two senses of smell or sensory channels: the standard sense that involves the olfactory organs, and a vomeronasal system similar to the snakes, in which receptors in the mouth detect chemical compounds and transfer the information via a duct to the vomeronasal organ in the nose; this then transmits it to the brain for analysis. Hearing is well developed in many species, especially those with large pinnae, such as the bilby and the long-nosed bandicoot.

Marsupial classification is based upon teeth and toes. If they have more than two incisor teeth in the lower jaw (most have four) and many teeth generally, they are called polyprotodonts and are carnivorous. If they have just one pair of incisors (usually incumbent) in their shortened lower jaw, and fewer teeth generally, they are diprotodonts and are vegetarians, primarily grazing and browsing herbivores. The diprotodonts occur only in Australia, whereas polyprotodonts live in Australia and South America. A further characteristic of the Australian diprotodont marsupials is the unusual structure of their feet, which are said to be syndactylous.[3] The second and third toes are fused together but with the claws separate. The fifth toe (the hallux or big toe) is missing, and the fourth toe is usually greatly enlarged and clawed—an adaptation for saltatorial or bounding locomotion.

American Marsupials

There are 85 species of American marsupials, most of which belong to the family *Didelphidae*. They have a wide range, from southeastern Canada through the United States and Central America down to Argentina, but only one species— the Virginian opossum—is native to North America, all the others occurring in the warmer regions of the New World. They are generally solitary animals and are nocturnal or crepuscular with one exception, the diurnal short-tailed opossum (*Monodelphis dimidiata*) of Argentina.

Unlike their descendants in Australia, many of which have become vegetarians, the American species are all polyprotodont animals, with dentition typical of the insectivores and carnivores. A few also eat fruit and are therefore omnivorous. Also, unlike their Australian diprotodont descendants, which lack a big toe, they are not missing any digits and have five on each foot; and in most species, to aid climbing their great toe is clawless and is used thumb-like in opposition to the other toes. The yapock (*Chironectes minimus*), which is semiaquatic, can close its pouch opening tightly when it enters the water.

The American marsupials are all small or medium-sized animals, ranging in size from the forty-six species of mouse opossums, which have an average head and body length of 6 inches (15 cm), to the common opossum, which may be 18 inches (45 cm) long. The smaller species resemble mice and rats, and the shrew opossums could easily be mistaken for real shrews (*Soricidae*), with their pointed snouts, tiny eyes, and poor vision. But their hearing and smell are well developed and the sense of touch also, as they have long, tactile whiskers. The New World marsupials are characterized by their elongated muzzles, small cranium, and long whiskers, and most have long, naked, or sparsely furred and scaly tails, rat-like except for their prehensile nature, being able to curl around objects and support the animal's weight. Shrew opossums (*Caenolestidae*) totally lack tail prehensibility, and the tails of the gray short-tailed opossum (*Monodelphis domestica*) and the thick-tailed opossum (*Lutreolina crassicaudata*) are only partially prehensile. The members of the genera *Didelphis*, *Chironectes*, and *Philander* have pouches, but in all other species the "pouch" is just an open area with folds of skin on either side of the nipples, and it cannot contain the young, so they must find a nipple, which then swells inside their mouth to secure them.

The New World marsupials have very short gestation periods, seldom more than twelve days, but even at that age the tiny embryos have well-developed arms and claws and clamber to the nipples without assistance from their mother, a rather precarious journey, especially if the mother is disturbed during the process. Like the young of the more familiar kangaroos and wallabies, baby opossums are initially cold-blooded and are totally dependent upon their mothers' body warmth to maintain their temperature. The babies, which grow in an enclosed pouch, are able to survive re-breathing air that has almost twenty times more carbon dioxide than fresh air. The American opossums have had many years to perfect the art of nocturnal living and are endowed with superb senses. They have large eyes and good night vision, and all have naked ears, which are quite large in the mouse opossum (*Marmosa incana*) and the Patagonian opossum (*Lestodelphys halli*) (the southern-most marsupial), and well-developed hearing. Their sense of smell is quite keen also, and assists the common opossum to locate garbage cans and dumps. Most species have long and tactile facial whiskers, which assist in the location and capture of insects.

Virginian or Common Opossum (*Didelphis virginianus*)

The only North American opossum, with a range from southeastern Canada to Central America, the Virginian opossum is the largest of the New World marsupials, with a head and body length of 20 inches (50 cm) when adult, and may weigh 11 pounds (5 kg). It is nocturnal, it is both terrestrial and arboreal, and it is a very good climber, aided by its sharp claws and powerful prehensile tail, which is as long as its head and body, and can support its weight. Its unusual, sparse-looking coat is grayish-black with long white-tipped guard hairs, and it has a long and pointed grayish-white face, with bare black-and-white ears. "Possums" are solitary animals, best known for their habit of "playing possum" or feigning death, a state of catatonia in which they are totally immobile. It is uncertain whether this can be controlled by the opossum or is totally uncontrollable, like fainting in humans. They are omnivores that hide in tree holes, burrows, or rock crevices during the day, and forage after dark for virtually anything edible, including invertebrates, amphibians, small reptiles, rodents, and birds and their eggs. They also eat carrion, fruit and other plant matter, and in turn are eaten by owls and by people in some regions. Northern opossums store fat in the fall and become inactive during very cold weather. Their well-developed pouch usually contains thirteen nipples but occasionally more, yet there is rarely enough for all the babies that are born, which is often more than twenty. When they find a teat they become attached as it swells in their mouth, so there is no sharing of nipples like piglets and puppies, and the excess young simply perish. Surprisingly, this is actually a safety measure, for if she produced just enough embryos for her nipples the high chances of losses would mean that nipples would be unused and wasted.

Murine or Mouse Opossums (*Marmota spp.*)

The murine opossums are a genus of about fifty species of small pouchless opossums from Central and South America, which vary in size from 3 to 7 inches (8–18 cm) in head and body length, and have long, thin and bare prehensile tails up to 10 inches (25 cm) long. Their fur is short and velvety and their coat color

Virginian Opossum *The only North American marsupial—and therefore the most primitive mammal on the continent—the common or Virginian opossum is a nocturnal tree-climber with a strong prehensile tail. Its pouch rarely has sufficient nipples for all of the babies that are born.*

Photo: Courtesy Harcourt Index

varies from gray-brown to dark reddish-brown, but they all have have paler faces with black markings around the eyes. The mouse opossums are shy and solitary forest animals, and with their nocturnal and arboreal habits are rarely seen. They are quite common, however, and can be easily caught in traps set at night and baited with ripe fruit. Although they are mainly carnivorous, with a wide diet that includes insects, lizards, small roosting birds, eggs or nestlings, and rodents, they also eat fruit and flower nectar and raid plantations, especially for bananas, paw-paws, and mangoes. They have occasionally travelled accidentally to North America in shipments of tropical fruit, especially in stems of bananas. The mouse opossums build nests in trees, on the ground beneath rocks, or in hollow logs, and they may use abandoned birds' nests.

Four-eyed Opossum (*Philander opossum*)

This is a slender marsupial from Central and South America, also known as the philander opossum, which reaches a length of 14 inches (36 cm) and has a prehensile tail 12 inches (30 cm) long. Adult males weigh just over 1 pound (900 g) and are slightly larger than the females. The four-eyed opossum has a short and dense, grayish-black coat with a darker head and a black-encircled white spot above each eye, hence its name. The tail is mostly naked and black at the base, with a pinkish tip. Four-eyed opossums are highly carnivorous and their natural diet includes insects, small mammals, birds, reptiles, amphibians, crustaceans, and snails. They may also eat ripe fruit, which should then in theory make them omnivorous, but the occasional vegetable matter consumed by an otherwise carnivorous animal does not really constitute omnivory. Like the mouse opossums, they are forest animals, usually found near water, and have both nocturnal and crepuscular habits. They build nests of grass and leaves, conical in shape, in a tree or an abandoned rodent burrow, to shelter in during the day, but they have pouches in which to carry their young.

Australian Marsupials

Despite their origins in South America, Australia is the marsupials' kingdom, with 195 species that have evolved into a far wider variety of animals than their ancestors, with both carnivorous (polyprotodont) and vegetarian (diprotodont) forms. They are the dominant mammals in Australia, and the presence of a few indigenous typical Old World species on the island continent, such as bats and rats, indicates restricted colonization by flying or seaborne invaders from the neighboring Indonesian Archipelago and New Guinea. All other wild mammals in Australia are the result of more recent deliberate or accidental introductions by human agency. The ancestral marsupials in America avoided the reptilian predators by being active at night, and many of the later evolving mammals elsewhere adopted nocturnal habits for the same reason or to avoid the sun. Predators then became nocturnal to take advantage of the source of food available at night. It is therefore surprising that the many new forms of marsupials, which radiated in Australia from the early possums, would retain their ancestors' nocturnal habits in the absence of native predators. Undoubtedly, species such as the red kangaroo preferred not to seek food under the intense outback sun, but in Tasmania's temperate climate the pademelon (a wallaby) has little need for such nocturnal behavior. One notable exception, active by day, is the numbat or banded anteater (*Myrmecobius fasciatus*), a marsupial that has evolved to fill the role of the anteaters and pangolins of the other continents.

When the first Australian marsupials radiated to fill vacant niches they independently developed similar characteristics to the evolving mammals elsewhere in the world, a process known as convergence or convergent evolution. Consequently, the Australian marsupials now resemble the flying squirrels, shrews, mice, anteaters, cats, dogs, rodents, and even moles of other continents, and like those animals they have evolved diets that are just as varied. They include terrestrial grazers and arboreal browsers, hunters, worm-eaters, honey-eaters, and scavengers. The large, grazing marsupials have developed a compartmental stomach similar to that of the higher ruminants (cattle, sheep, and goats) for digesting their fibrous diets by microbial fermentation. Unlike their insectivorous or carnivorous ancestors the Australian marsupials diverged to take advantage of plant matter in its various forms—grass, leaves, fruit, nuts, flowers, and honey—and developed mouths, teeth, tongues, and digestive systems to cope with their vegetarian lifestyle. These diprotodont marsupials include the wombats, koalas, kangaroos and wallabies, and many species of possums. The polyprotodont or carnivorous marsupials include the marsupial mice, native cats, and the bandicoots.

Dasyures

In the absence of the true carnivores evolving in the rest of the world, some Australian marsupials took advantage during their own development of a plentiful source of food—other marsupials—and became a family of meat-eaters. The dasyures, of the family *Dasyuridae*, fill the role of the cats, dogs, and smaller hunters of other lands. They not only resemble many members of the *Carnivora* externally, but are anatomically and physiologically adapted for consuming their prey, with large and well-developed canines with sharp cutting edges and sharp, cusped molars, plus a simple meat-digesting gut rather than the bacterial-fermenting tracts of the herbivorous marsupials. The pouch is absent in some species, and poorly developed and barely discernable in others, except in the breeding season when it may just be folds of skin surrounding the nipples. This is compensated for by the evolutionary adaptation of the teat swelling in the baby's mouth, firmly anchoring it, allowing it to dangle freely as the mother runs along the branches. Dasyures range in size from one of the world's smallest mammals, the tiny flat-skulled marsupial mouse (*Planigale tenuirostris*), which weighs just $\frac{1}{6}$ ounce (5 g) to the Tasmanian devil (*Sarcophilus harrisi*), which reaches 22 pounds (10 kg) in weight, and, until recently, to the now-extinct dingo-sized Tasmanian wolf.

With the exception of the mostly scavenging Tasmanian devil, in keeping with their predatory lifestyle the dasyures are active and agile animals with well-developed senses of sight, hearing, and smell. They are aggressive predators, with rapid reflexes and swift movement, which rely on a range of prey from insects for the marsupial mice to animals as large as wallabies for the native cats, which may also eat fruit and other plant material.

Eastern Native Cat or Quoll (*Dasyurus viverrinus*)

A smaller animal than the similar tiger cat, with a shorter and more bushy tail, this species has two color forms—black and olive-brown—both being heavily spotted with white, but on the body only, unlike the tiger cat (*D. maculatus*), which has a spotted tail. The quoll is found in southeastern Australia and Tasmania in dry schlerophyll forest with rocky areas, where it is mostly terrestrial. It is about the size of a pine marten, with head and body length of 18 inches (45 cm) and a 12-inch-(30 cm)-long tail. An adult male weighs about 3¼ pounds (1.5 kg) and the female 2 pounds 3 ounces (1 kg). Now a rare animal in mainland Australia, it continues to thrive in Tasmania. The native cats are the equivalent of the civet "cats" of Africa and Asia rather than the true cats. They are active at night in their search for small mammals and birds, and are known to leap from trees to catch passing birds in flight. They can fold their ears down during the day to block out sounds. Like the true carnivores their teeth are adapted for catching and tearing. Their pouches are little more than two folds of skin surrounding their nipples. Quolls have granular or smooth palm and sole pads indicating a basically terrestrial way of life. They are well-adapted nocturnal hunters in which no one sense appears to be dominant. They have large ears and good hearing, and although their eyes are not excessively large they provide good night vision; their sense of smell is excellent and they have large, tactile whiskers.

Tasmanian Devil (*Sarcophilus harrisi*)

This is certainly the most well-known carnivorous marsupial, a small, bear-like animal with a short and broad head, powerful jaws and heavy, bone-crunching molar teeth reminiscent of the hyenas. It survives in Tasmania only, having long been extinct on mainland Australia, possibly as a result of the introduction of the dingo. It favors coastal heaths and schlerophyll forest, and up to 4,000 feet (1,220 m) in the very wet forest of Cradle Mountain National Park. Devils are black with a white throat patch and occasionally have a white spot on the side or rump, and have pinkish-white snouts. Adult males have a head and body length of 26 inches (65 cm), a tail about 10 inches (25 cm) long, and weigh up to 24 pounds (10.9 kg); females are smaller and weigh only about 18 pounds (8.1 kg). They are noisy, belligerent, and antisocial animals, but hand-raised babies have remained very tame and handleable. However, they are certainly solitary, although nonterritorial, and do not appear to like each others' company. Even the act of mating is a long, very noisy, and tiring tussle, and the maturing young so pester their mother that she just walks off and leaves them as soon as they are weaned. Devils are nocturnal and terrestrial, and shelter by day in caves, rock crevices, wombat burrows, or hollow logs. With their well-developed sense of smell and foraging with nose to the ground they are efficient scavengers, particularly of Tasmania's largest wild mammals—kangaroos and wallabies—and of domestic sheep. Their population plummeted some years ago when they were wrongly accused of livestock predation, but they are now protected and have recovered. While they are certainly able to kill penned sheep, devils could never emulate the extinct thylacine and run one down. They apparently once relied largely on the remains of thylacine kills, but have managed quite well since its demise.

Bandicoots and Bilbies

These unusual Australian marsupials belong to the families *Peramelidae* (the bandicoots) and *Thylacomyidae* (the bilbies or rabbit-eared bandicoots). They slightly resemble large kangaroo rats, but have the dentition of the placental insectivores and the larger hindlimbs, long hind foot, and syndactylous toes of the terrestrial and herbivorous kangaroos, but they differ in having long and sharp claws on their forefeet for digging. They also lack the kangaroo's procumbent lower jaw incisors, having instead tiny incisors that reflect their mainly insectivorous diet, and their pouches open to the rear like the wombats and koalas. Both the bilbies and bandicoots have large ears and nonprehensile, rat-like tails; their long and pointed muzzles are very flexible and are used for rooting in the soil for invertebrates. They are nocturnal and terrestrial and spend the daylight hours either in a nest of twigs and leaves or in a burrow of their own digging. Together these species form the second group of predatory Australian marsupials, and although they are basically insectivorous, their main foods being insect larvae, earthworms, and beetles, they are also aggressive predators of lizards and small mammals. They have good hearing and a good sense of smell, but their vision is poor.

Long-nosed Bandicoot (*Perameles nasuta*)

This bandicoot lives in the rain forests and sclerophyll[4] forests of eastern Queensland, New South Wales, and Victoria. It has a variable coat color, the most common shades being light brown, dull orange-brown, and gray-brown, and is about 15 inches (38 cm) long with a short, rat-like tail and large, bare upright ears, which it can fold over when asleep. Like all four species of the bandicoots with long noses, it is terrestrial and highly nocturnal, and spends the daylight hours in a nest of twigs and leaves on the ground in heavy bush; but in open areas may dig a nest hole, line it with grass, and then cover it with leaves and twigs. It also uses the abandoned burrows of other animals, hollow logs, and rock piles. It is a very territorial animal and males are quite aggressive to intruding conspecifics. The pouch is well developed and has eight mammae, but litters rarely number more than four babies, so that the next litter can find unused nipples, for they could not attach themselves to the just-vacated, swollen ones. The long-nosed bandicoot has excellent hearing and a keen sense of smell but its eyesight is poor. It digs and scratches in the undergrowth for worms, snails, and other invertebrates—opening a hole with its long claws into which it can push its snout and sniff out its food.

Rabbit Bandicoot or Bilby (*Macrotis lagotis*)

Bilbies have huge, naked ears, which they fold flat when sleeping, and their sense of hearing is well developed; they also have a good sense of smell, but not very good eyesight. They have a head and body length of 20 inches (50 cm), and long pointed snouts. Ashy-gray or fawn-gray in color with paler underparts, their 10-inch-(25 cm)-long tails are black at the base and white on the terminal half. Bilbies are terrestrial and are the most nocturnal of the bandicoots, which spend the daylight hours in burrows dug with their powerful limbs and foreclaws, loosening the soil by scratching with their forepaws and then kicking it backward with the hind legs. Their burrows are unusual, steep, and spiralling, and up to 6 feet (1.8 m) deep so that they escape the heat of the day, for they live in the dry country—semidesert, grassland, and dry woodland—of central and southern Australia. They hop on their hind feet like the kangaroos and also shuffle along on all fours. Two young are born after a gestation period of fourteen days, and spend about eighty days in the backward-facing pouch. Bilbies were formerly included with the bandicoots in the family *Peramelidae*, but were recently reclassified *Thylacomyidae* due to differences in their skulls and dentition.

Phalangers

The phalangers are a group of nocturnal and arboreal marsupials that are primarily leaf and fruit eaters, and are therefore diprotodont animals. They belong to the families *Phalangeridae*, *Burramyidae*, and *Petauridae*, and include the phalangers, cuscuses, possums, ringtails, pygmy possums, and the gliding phalangers or sugar gliders. In keeping with their arboreal lifestyle phalangers' tails are prehensile and they have a well-developed pouch that opens to the front. They were originally called opossums by Captain Cook because of their similarity to the American marsupials, but this was later changed to possums to avoid confusion.

The ring-tailed possums and the cuscuses are the two largest groups, and are truly arboreal and nocturnal animals with strong prehensile tails. The possums are nest-builders and shelter in these by day; when active at night they are thought to rely equally on their vision and smell, the youngsters apparently being able to follow trails made by the parents, and their large, upstanding ears indicate a good sense of hearing. The cuscuses are densely furred marsupials that bear a striking resemblance to the kinkajou in form and habits, with round heads, tiny ears and large eyes, and a prehensile tail, except the cuscus's tapers at the end and is bare on the terminal third. They differ also in their diet, however, as cuscuses are mainly folivores, whereas the kinkajous are frugivores. In keeping with their leafy diet they have large teeth for fiber mastication, a simple stomach, and a large blind appendage or caecum in which the leaves are fermented. Cuscuses have one of the widest ranges of the Australian marsupials, including northeastern Australia, New Guinea, and many islands from Sulawesi to the Solomon Islands, and their flesh is eaten by the natives of some islands. The flying phalangers or sugar gliders of the family *Petauridae* are totally arboreal and nocturnal animals, but their tails are not strongly prehensile. They are omnivorous, favoring a diet of fruit, nuts, and invertebrates, but like the marmosets of the neotropics they rely to a large extent upon plant gums. Gliders are social animals that normally live in groups of an adult male, several females, and their young.

Brush-tailed Possum (*Trichosurus vulpeculus*)

A common animal throughout most of Australia, except the northern parts of the Northern Territory, this possum is solitary, nocturnal, and mainly arboreal, and is a very specialized folivore or leaf-eater. Yet it also occurs in semidesert regions where trees are sparse, and where it lives in caves and holes; and it has colonized urban areas where it is Australia's equivalent of the raccoon, raiding garbage cans and dumps. It has large, bare ears and a prominent pink nose, and adults measure up to 20 inches (50 cm) long, have a 12-inch-(30 cm)-long tail and can weigh up to 9 pounds (4 kg) (see color insert). This possum has a thick and woolly coat which is usually silvery-gray, with a black end to its long and bushy prehensile tail, but there are several color phases, including an all-black one in Tasmania. Its pelt was in demand by the fur industry and to increase the supply it was introduced into New Zealand in 1837. It responded with such rapid colonization that 400,000 pelts were harvested in 1945. But like so many alien animal introductions this was not without its cost to the environment, for possums have extensively damaged pine plantations and completely defoliated and killed native trees that had evolved in the absence of herbivorous mammals. They eat birds' eggs and young, including kiwis, and in some districts have affected the honey crop by extensively damaging the rata blossoms.

Spotted Cuscus (*Phalanger maculatus*)

The spotted cuscus is an animal of the rain forest and mangroves of Australia's Cape York, New Guinea, and its neighboring islands. Despite its name it is very variable in color, with one of the animal kingdom's most unusual arrangements of coat color and pattern. Adult males are normally creamy-white with lots of black or rufus spots and blotches, but they may be plain colored, or almost white with a few spots. Adult females are usually grayish-brown and unspotted, except on the island of Waigeo where the sexes are alike—both having a creamy ground color with a reddish hue and lots of spots. During their growth stages the cuscuses' color and pattern may also change. The spotted cuscus has a head and body length of 18 inches (46 cm) and weighs 7½ pounds (3.4 kg). Its prehensile tail, which is almost as long as its body, is furred on top for about three-quarters of its length. Its eyes are very large and yellow-rimmed, and have vertically slit pupils in contraction. Although highly arboreal the cuscus does come down to the ground, when it bounds along with surprising speed; but generally it is a shy and solitary animal with a rather slow and sleepy disposition. It sleeps during the day in dense foliage where it is easily caught by the natives and their dogs.

Sugar Glider (*Petaurus breviceps*)

Sugar gliders have soft, pale-gray coats with a dark-brown stripe down the back and yellowish-white underparts. Their head and body length is about 7 inches (18 cm), and they have bushy tails, which are partially prehensile. Females are usually a little smaller than the males when adult. They are very social animals, with loud voices that carry for several hundred yards when they are gliding, and they growl out of all proportion to their size when disturbed in their nest. Their evolution in Australia and New Guinea and consequent adaptations have resulted in a perfect example of convergence, where these small marsupials now resemble the North American flying squirrels in form and behavior. Like the rodents they emulated, the sugar gliders have large gliding membranes and are colonial and nocturnal, sleeping all day in a nest of leaves in a hollow tree. Their membranes, which lie between the forefoot and ankle and open when the limbs are spread, can support them for downward glides of 160 feet (49 m) between trees. They are omnivores, and eat insects and a wide range of vegetable matter such as fruit, nuts, shoots, and plant sap.

Mountain Pygmy Possum (*Burramys parvus*)

Although it is the largest of the pygmy possums this marsupial is a tiny animal, weighing only 1½ ounces (42 g). It is highly secretive and is strictly nocturnal, sleeping by day in a nest among the rocks in its very small habitat on the mountain top heathlands in the Victoria Alps of southeastern Australia. Now a very rare animal, it was in fact believed extinct until rediscovered in 1956 on Mt. Higginbotham and then in 1996 on neighboring Mt. Buller, but its small population must now cope with alien predators such as foxes and feral cats and dogs, plus human activities such as mountain biking, trekking, and skiing. The mountain pygmy possum is the only truly alpine marsupial and the only species to have a long seasonal hibernation, becoming torpid in March or April (the southern autumn) in a nest in a deep crevice or among rocks. Over winter, the nest is covered with at least 3 feet 3 inches (1 m) of snow, which insulates the possums from the subfreezing temperatures of the mountains. The nest remains constant at 35°F (2°C), but the possum arouses periodically to move around in the rocks beneath the snow. It sleeps until October or November, and breeds when it emerges from hibernation.

Rat Kangaroos

These unusual animals resemble small, smooth-coated wallabies, with similar pointed muzzles, long syndactylous hind feet, and a form of saltatorial locomotion, although with a more "hunched" hopping gait with their tail held out straight, and using both their front and back limbs when they run, which resembles a gallop. Unlike the other kangaroos their front limbs are adapted for digging, for their diet consists primarily of roots and insects, and they cannot kick with their hind feet like the big kangaroos. They are considered a more primitive group of marsupials than the kangaroos and are included in their own subfamily, the *Potoroinae*. Rat kangaroos are solitary, nocturnal animals with well-developed pouches like the other macropods, but unlike them they feed in dense cover rather than out in the open. Several species carry nesting material with their tails which are semi-prehensile, an unusual adaptation for ground-dwelling marsupials, or in fact for any terrestrial mammal.

Long-nosed Potoroo (*Potorous tridactylus*)

This species has a discontinuous distribution, occurring in the southwestern corner of Western Australia and then in the east, from southern Queensland down into Victoria, and not in the arid country in between. It lives in tussock grassland, heaths, and open eucalypt forest, wherever there is sufficient tall grass cover, and being strictly nocturnal sleeps during daylight in depressions beneath vegetation or in a nest made at the base of a tree, but can occasionally be seen sunbathing. It is grayish-brown above and pale gray below, with a head and body length up to 16 inches (40 cm), a sparsely furred tail 10 inches (25 cm) long, and an adult weight of 4 pounds (1.8 kg). The potoroo has a long and pointed muzzle similar to the bandicoots, and although the hind feet are short it makes very long jumps for such a small animal, and at high speed can clear 11 feet (3.3 m) in one bound at a height of 5 feet (1.5 m). It is highly insectivorous, but also eats roots and grass and has a special liking for fungi.

Brush-tailed Bettong (*Bettongia penicillata*)

The brush-tailed bettong formerly occurred over a vast area of Australia, in the southwest and southeast corners, and was a very abundant animal in the mid-nineteenth century. It is now one of the rarest marsupials, extinct over much of its range since the 1920s, and now surviving only in six sites in Western Australia and on three islands and two mainland sites in South Australia, after the involvement of captive breeding, reintroductions, and the control of foxes. Its decline has been attributed to loss of habitat to grazing animals and clearance for agriculture, and predation by the introduced red fox. Brownish-gray above, with pale underparts, it has a ridge of long hair along the lower spine, and a head and body length of

14 inches (35 cm) and a weight of 3¼ pounds (1.5 kg). Its tail, which is about 12 inches (30 cm) long, is lightly furred except for the end third which is covered with bushy black hair.

The bettong lives in dry grassland, heaths, and dry woodland, where it makes a nest of grass on the surface, carrying the grass in the typical bettong fashion in its curled tail. It is an omnivore and extends its typical kangaroo diet of shoots, roots, and seeds with invertebrates, but it is especially fond of fungi, which may even account for the bulk of its diet. However, just as mammals cannot digest plant fiber without the assistance of bacterial fermentation, the bettong is assisted by bacteria that digest the fungi and produce nutrients that the bettong can in turn absorb in its small intestine.

Kangaroos and Wallabies

These are the largest marsupials, which belong to the family *Macropodidae* and are therefore often called macropods. In Australasia they occupy the niche of the grazing and browsing ruminants (deer, antelope, cattle) of other lands, bacterially fermenting their fibrous diet in their compartmentalized stomachs similar to the placental ruminants, a process known as pre-gastric fermentation. There are fifty-two species of these distinctive animals, characterized by their small heads and large ears, heavy hindquarters, and long, muscular tails, which act as a prop when they are resting, as a lever when they rear up to fight, and a counterweight when they move fast. The other unusual feature of these macropods, and several other species such as the bandicoots, is the syndactylous structure of their hind feet. They have five toes on the forefeet, but appear to have only three on the hind feet, as the first toe is missing and the two innermost ones are enclosed in a common sheath from which two nails protrude and are used as a fur comb. The center toe—which corresponds to the fourth toe in the human foot—is greatly enlarged, with a foot pad from heel to toe, and is tipped with a heavy nail. The terrestrial kangaroos and wallabies employ the remarkable saltatorial or bipedal bounding mode of locomotion on their sturdy hindlimbs, the larger species achieving speeds up to 30 mph (48 kph) and able to clear up to 13 feet (4 m) horizontally with each bound. Excluding the mainly diurnal whiptail wallaby (*Macropus parryi*), the kangaroos and wallabies are nocturnal or crepuscular, although they may occasionally appear during daylight, and several species do like to sunbathe. The forest and bush species rest in dense cover during the day and are rarely seen; the rock wallabies and wallaroos sleep on rock ledges sheltered by overhanging rocks; and the red kangaroo of the semidesert outback finds shade from the fierce sun wherever it can, usually under bushes in open view.

Red Kangaroo (*Megaleia rufus*)

The red kangaroo is the most familiar of the large macropods and is the species most frequently seen in zoos. It is the largest marsupial and has been the mainstay of the hide industry and more recently of the meat industry for many years, with several hundred thousand animals killed annually for commerce, yet the exportation of live animals is prohibited except to approved noncommercial zoological gardens. The sexes are quite distinct: males have bright rufus upperparts which in the females are slaty-gray with a pink tone; the underparts are pale grayish-white and the ears gray or brown in both sexes. The fur on their backs and sides is close and woolly, composed almost entirely of underfur. The red kangaroo is a robust animal, in which adult males have a head and body length of 5 feet 3 inches (1.6 m) and a tail 4 feet (1.2 m) long, and can weigh 132 pounds (60 kg). When standing upright and balancing on their thick tails ready to kick out with their hindlimbs, they can look a person of average height in the eye, and are always ready to use their feet or their teeth when upset or manhandled. This species is an animal of the open timbered plains, grasslands, and semiarid regions of central Australia, which has also benefitted from the changes in vegetation resulting from the development of sheep pasture and artesian wells.

Red Kangaroo   *Primarily nocturnal to avoid the intense midsummer heat of the Australian outback, red kangaroos do like to sunbathe for short periods. The largest kangaroos, and the largest marsupials, they are still killed in the thousands annually for their meat and hides.*
Photo: Courtesy Ralf Schmode

Eastern or Great Gray Kangaroo (*Macropus g. giganteus*)

A forest and open woodland species from Queensland, New South Wales, Victoria, and Tasmania, the great gray kangaroo is a large and robust animal, apparently the first kangaroo seen by Captain Cook in 1770. It has a head and body length of 60 inches (1.5 m), with a tail measuring 36 inches (92 cm) in length; its fur is short, close, and woolly, silvery-gray above, with underparts and limbs grayish-white and a brown tail with a darker tip. It is a grazer and a browser, and although all the saltatorial marsupials are believed to swim (but seldom have the opportunity), the great gray is a strong swimmer that has been seen over 1 mile (1.6 km) from the shore. It has increased in numbers as a result of sheep farming and the availability of grazing land and stock watering points, and is still heavily hunted for its meat and hides. There are three races, the western one known as the Mallee kangaroo (*M. g. melanops*) has a black patch on top of the muzzle and between the eyes, and is also known as the black-faced kangaroo. The race in Tasmania, *M. g. fuliginosus*, has a longer and coarser coat, befitting an animal that has evolved in a cool and damp climate.

Red-necked Wallaby (*Macropus rufogriseus*)

The red-necked wallaby is a forest and woodland species from eastern Australia and the island of Tasmania. On the mainland it lives mostly to the east of the Great Dividing Range and on its slopes, but in a few places it has gone over the mountains into the woods on the western side. Adult males weigh up to 44 pounds (20 kg) and females up to 26 pounds (11.8 kg), and the largest specimens have a head and body length of just over 40 inches (1 m) and a tail 30 inches (76 cm) long. It is the most familiar wallaby in zoos, and the one often kept as a free-ranging animal on northern country estates is the race known as Bennett's wallaby (*M. r. fructicus*), from Tasmania and islands in the Bass Strait, which has longer and coarser hair (an adaptation for the cooler and wetter climate) and is therefore predisposed to cool, temperate northern environments. It has more somber coloration than the typical form, with a neck and rump of dull grayish-rufus, and an almost black back and grayish-white underparts. The race known as the red-necked wallaby (*M. r. banksianus*) is a grayish-fawn animal in which the back of the neck and rump are bright rufus. Thousands were killed annually for their skins during the early years of colonization of southeastern Australia, but it has benefitted from forest clearance and the development of farmland; in some areas it is now more abundant than when settlement began and is considered a pest.

White-fronted or Parma Wallaby (*Macropus parma*)

A small species that weighs 11 pounds (5.5 kg), the parma wallaby has a rufus back and a distinct white cheek stripe and dark streak on the back of its neck. But its most distinguishing feature is its pure white throat which contrasts with the gray sides of the neck and its grayish-white underparts. The parma wallaby was believed extinct in its natural range of just two small areas of eastern New South Wales early in the twentieth century due to loss of habitat and predation by introduced red foxes; but in 1966 a small population was found on Kawau Island in New Zealand's Hauraki Gulf, where they had been introduced by the prime minister Sir George Grey in 1870, during the age when many alien animals entered New Zealand. Since then survivors have also been rediscovered in its original Australian range, and a management plan has been developed for their protection. The parma's survival has also been aided by the animals that were exported from New Zealand to many zoos, where captive breeding has been excellent and has further aided the wallaby's recovery.

Matschie's Tree Kangaroo (*Dendrolagus matschiei*)

Adapted for arboreal life, the tree kangaroos have powerful forearms for climbing, long and well-furred cylindrical tails which they use for balance, and short and broad hind feet with curved nails for gripping branches. They have stocky bodies but their hindlimbs are not as well developed as those of the bounding terrestrial species. Mature animals are about 32 inches (81 cm) long and can weigh up to 22 pounds (10 kg). The tree kangaroos have been called agile climbers, but that depends on which way they are going. They can climb straight up a tree trunk quite swiftly, hugging it in the manner of a lumberjack using a climbing rope and irons, but they come down tail first, and that is a very laborious business. But they are good jumpers, from branch to branch and especially downward to other branches, and they can jump safely to the ground from a height of 40 feet (12 m). They are quite social animals and tend to form groups of a male and several females in the wild; they are mainly nocturnal in their habits. Their gestation period of about forty-two days is the longest of all the marsupials. Matschie's tree kangaroo hails from New Guinea's Huon Peninsula, plus neighboring islands where it was believed to have been introduced by Melanesian traders long ago. It has a lovely thick and soft coat that is reddish-brown on the back, with face, feet, belly, and tail buffy-yellow. It is the most colorful of the macropods (see color insert).

Koala

A single species in the family *Phascolarctidae*, the koala (*Phascolarctos cinereus*) is one of the world's most recognizable animals. A stocky and tail-less marsupial, it has a head and body length of 33 inches (84 cm); adult males may weigh up to 22 pounds (10 kg) and females 18 pounds (8.1 kg). The koala has a massive head, a large nose and cheek pouches, large incisor teeth, tiny canines, and well-developed molars for masticating leaf fiber. It has a dense, woolly coat, light gray above and white below; and large ears fringed with white. Totally arboreal, the koala has strong claws and granular soles on its feet and hands to aid climbing, and its pouch opens to the rear—unusual for an arboreal animal. A very specialized folivore, it eats mostly the leaves and tender bark of just a few species of eucalypts, and unlike the ruminating kangaroos and wallabies, which breakdown their herbage in a specially compartmentalized stomach, the koala has a very long cecum (up to 8 feet [2.4 m] long) for bacterially fermenting leaves. When her baby is ready for weaning at the age of about six months, the mother koala feeds it digested leaves from her anus, inoculating the youngster's intestine with the essential symbiotic bacteria.

Koala *A strict folivore, or leaf-eater, the koala rests in the fork of a tree during the day. Active at night, it eats only the leaves and bark of a few species of eucalyptus, bacterially fermenting them in its very long cecum prior to digestion taking place.*
Photo: Clive Roots

The koala is mainly solitary and territorial—at least in the tree it occupies—and a male attempts to maintain control over several females during the breeding season, which is generally during the Australian summer from September to February. Keeping koalas outside Australia has always been uncommon, due to strict protective measures and the need for regular supplies of fresh eucalyptus. Permanent displays of koalas have been restricted to zoos in zones where eucalypts grow, such as California and Japan.

Wombats

Wombats are large, stockily built animals with short but very powerful limbs and strong claws, massive heads, large noses, and small eyes; like the koala their pouches open to the rear. Their distribution includes some of the hottest and driest regions of Australia, where they survive by virtue of their behavioral and physiological adaptations. They do not sweat and are therefore vulnerable to heatstroke and must shelter from intense heat; they also live in the coldest regions of Australia, and are the only large burrowing marsupials, which allows them to escape extreme conditions in burrows that may be 6 feet (1.8 m) deep and perhaps 35 feet (10.6 m) long. In the drier parts of their range wombats conserve water by concentrating their indigestible food matter into very dry fecal pellets, and like the kangaroos can survive for long periods without drinking. Wombats have small eyes and small ears and, unlike the other diprotodont marsupials, which have three pairs of upper incisors, the wombat has just a single pair to oppose the lower pair, like the rodents. These teeth are rootless and grow continually, so they are chisel-like with an ever-sharp cutting edge, and they have enamel only on the front and lateral surfaces, again like the rodents, another example of marsupial convergence.

Common or Forest Wombat (*Vombatus ursinus*)

Common by both name and numbers, this species is a resident of the rocky, schlerophyll forest of eastern Australia and Tasmania, and is considered reasonably safe as its habitat is not required for agriculture. It has a very coarse coat of thick brown fur, short ears, and bare granular skin on the nose; it is the larger of the three species, about 48 inches (1.2 m) long, with just a stub of a tail and a weight of 75 pounds (34 kg).

Southern Hairy-nosed or Plains Wombat (*Lasiorhinus latifrons*)

This hairy-nosed wombat[5] lives in the coastal regions of southern Australia from the Murray River west across the Nullarbor Plain. It differs from the previous species in its large ears, fine silky coat, and fine hairs on the nose. It is also smaller, reaching a length of just over 36 inches (92 cm) and a weight of about 65 pounds (29.5 kg), and it digs very extensive burrow systems to escape the region's intense summer heat.

Notes

1. The generation or production of heat, especially by physiological processes.
2. Except for several species that have extended their range into Central America, and the common opossum, which has colonized North America up to southern Canada.

3. The bandicoots, which are polyprotodonts, also have syndactylous front feet.

4. Typically Australian forests in which the trees have hard and short evergreen leaves.

5. There is also a northern hairy-nosed wombat, a very rare animal (the rarest marsupial) of which perhaps only twenty survive in Queensland, Australia.

6 The Sound Navigators

Fossils prove that the bats of 50 million years ago looked very similar to those alive today, which means they are a very successful group of animals, even though the ancestors of those early bats have yet to be determined. The modern bats form the second-largest group of mammals after the rodents, and belong to the order *Chiroptera*, which is divided into two suborders. *Megachiroptera* contains the world's largest species, the Old World fruit bats, all in the family *Pteropodidae*. The other suborder, *Microchiroptera*, contains all the other living bats, about 800 species in 16 families, which are usually referred to as insectivorous bats, although they are not all insect-eaters. Many have diverged from the basic insect-eating habits of the family to an omnivorous diet and even to carnivory.

Northern bats are said to be heterothermic, as their body temperature fluctuates in accordance with the ambient temperature. They cannot keep warm as winter approaches, and with food supplies dwindling they are forced to migrate or hibernate to escape the freezing temperatures. During hibernation or long-term torpidity their body temperature drops to that of their environment, which must therefore stay just above freezing; their respiration slows and their oxygen consumption is a fraction of its normal rate. Daily torpidity to conserve energy also occurs in the northern bats while they are sleeping, irrespective of the air temperature.

Tropical bats are more homeothermic, unlikely to be exposed to very low temperatures and therefore able to maintain their body temperature, and food is generally available year round so there is no need to migrate, but even tropical bats may become torpid if food is scarce, irrespective of the temperature. These bats are not restricted to insects and have the advantage of being able to specialize, and their numbers now include fruit-eaters, frog-eaters, and fish-eaters.

■ ECHOLOCATION AND THE SENSES

Most bats are truly nocturnal and rely on senses that allow them to fly at night or in the total blackness of caves where sight is useless; but a few are crepuscular and feed in the early morning and again at dusk before it becomes quite dark, and vision obviously plays a more important role in these species. Bats are amazing animals, with the most well-developed voice and sense of hearing of all mammals, able to avoid obstacles while flying at speed in total darkness, navigating by the echoes returned from their calls. How they do this was not clearly understood until quite recently. At the end of the eighteenth century the Italian scientist Spallanzoni proved that bats had to be in full possession of their vocal chords and their sense of hearing to avoid obstacles, but it was not until the middle of the last century, through the efforts of Dr. Donald Griffin and his colleagues, that it could be said with certainty that bats navigated in the dark by echolocation. They make ultra-sonic calls with frequencies of up to 100,000 vibrations per second, and locate obstacles by interpreting the returning echoes. In conjunction with this ability they obviously have exceptional hearing and rapid motor response. Early in the last century echolocation was investigated as a military aid, and as a result "sound navigation and ranging" (sonar) was developed during World War I, primarily for the detection of underwater objects, especially submarines.

In addition to avoiding obstacles by echolocation, bats also use their "kines-thetic" or muscle sense, which allows animals generally to move about in darkness in familiar surroundings. During experiments, bats that were accustomed to fixed objects in their area bumped into newly introduced or moved objects until they became used to their whereabouts. Other experiments proved that they are more likely to collide with objects if they are tired, or if they have just awoken and are not fully alert.

Animals dependent upon echolocation for their survival must be able to send out signals very fast. When they first take flight the bats' signals are emitted at the rate of about ten per second, but this builds up to fifty per second when they maneuver around obstacles. Experiments showed that navigating difficult obstacles resulted in an increase in their signals to 200 per second in short bursts. As the duration of these signals ranges from one-hundredth to one-thousandth of a second, the speed with which the returning echos are received, transmitted to the brain, their information analyzed, and the decision flashed to their wing muscles for action is truly amazing. But not all species utter supersonic calls. Some members of the *Phyllostomidae*—the tropical American fruit bats, which are not as dependent upon echolocation—utter short-duration, soft-sound impulses and are called whispering bats. Their calls have only a fraction of the sound energy of the insect-eaters. Among the Old World fruit bats only those of the genus *Rousettus* are known to produce supersonic calls, in addition to depending on vision for locating fruit. The ten species of rousette bats of Africa, southern Asia, and Indonesia have a buzzing call which is believed to be a simple echolocation system, but is not as developed as that of the insectivorous bats.

Bats emit their sounds through the mouth or the nostrils, depending upon the species, and many have elaborate nose leafs which orient their calls. These travel in a conical beam, which varies in diameter according to the species and the shape of its mouth and nose leaf, and the sound pulses are increased by some bats as they approach their prey. Bats' calls differ according to their habits; some have ultrasonic sounds with a single pitch,[1] but the North American insectivorous species can vary the pitch of their calls and are called frequency-modulating bats. The nose appendages or "leafs" take a variety of weird shapes, which include leaves, spikes, and in several species an elongated "spear." Coupled with the un-usual features of many, such as the wrinkle-faced bat (*Centurio senex*), whose face is as wrinkled as a Sharpei dog's; the self-descriptive mastiff bat (*Eumops perotis*) and the hammerhead bat (*Hypsignathus monstrosus*); and the Colima long-nosed bat (*Musonycterus harrisoni*), which has an extremely long snout with an upright spike on the end; it is hardly surprising they appear so sinister. Most bats are silent to our ears, their ultrasonic sounds far above the the human range of hearing, and in addition to their echolocating calls made during flight they also use high-frequency sounds when they are resting, presumably for communication in the colony. Some species also make sounds that we can hear. The larger fruit bats have well-developed larynxes, and roosting and feeding times are accompanied by loud squabbling. One member of this group, the hammerhead bat, has a greatly en-larged larynx that almost fills its chest cavity, and the heart and lungs are pushed backward and to one side to make room for it. Their deep-throated calls have been likened to a pond full of noisy frogs. The epauletted fruit bats (*Epomophorus*) and the tube-nosed fruit bats (*Nyctimene*) have whistling calls which can be clearly heard by the human ear.

Echolocation not only requires well-developed voices; it demands a sense of hearing that is developed in the bats to a far greater degree than in any other mammal. From the speed of the bat's calls and the rapidity of the action taken, it is obvious that in the insectivorous and carnivorous species at least, the sense of hearing is so acute they can receive and interpret echoes while making calls. Allied to their use of ultrasonics their complex ears have an additional hearing aid called the tragus, which is a small, cartilaginous flap in front of the external opening of the ear that is absent in the other, nonecholocating bats. The tragus varies in shape, being erect and tapering in the North American little brown bats and blunt-ended in the pipistrelles. Large ears are typical of many bats, being especially huge in the false vampire bats (*Vampyrum*), the yellow-winged bat (*Lavia frons*), and the spotted bat (*Euderma maculatum*), which has the largest ears of all North American bats. Bats with long ears can usually curl them back like ram's horns, or tuck them under their wings, when resting.

Bats can hear frequencies up to 100,000Hz[2] or cycles per second (cps), and they can distinguish between obstacles and their prey, even when flying in woodland where there could be interference from many much stronger echoes. It is interesting that the bats that "fly blind" by echolocation fly much faster than the fruit bats that fly by vision—proof of the effectiveness of the echolocation sys-tem. Dissection of bats' brains has revealed some interesting but not altogether

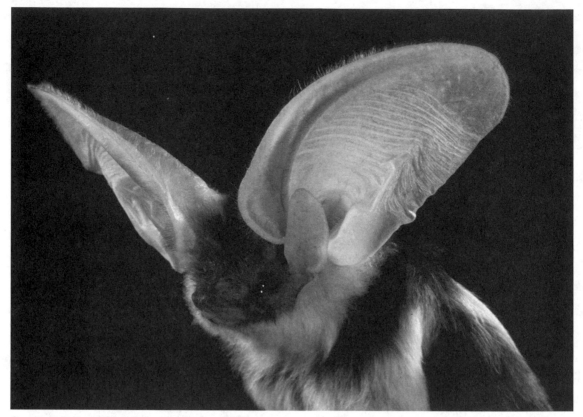

Spotted Bat   *With the largest ears of all North American bats, the spotted bat has acute hearing essential for the immediate receipt of echoes from its high-frequency calls, allowing it to navigate at full flight in complete darkness.*
Photo: Merlin D. Tuttle, Bat Conservation International

surprising details. In the *Megachiroptera* (mostly large fruit-eaters) the areas of the brain dealing with vision and hearing are similar in size, whereas in the *Microchiroptera* (mostly insectivorous species) the auditory areas are extensive and the areas concerned with vision are reduced.

All bats have functional eyes, and the saying "as blind as a bat" is incorrect. They can all see and the fruit bats (*Pteropodidae*) have very good vision, with many species relying more upon sight than echolocation; but the eyes of other bats are quite small, and are often obscured by fur. However, vampire bats have no difficulty scuttling across the floor to their intended victim, and captive insectivorous bats easily locate a dish of mealworms on their cage floor. Apart from the *Rousettus* bats, which use echolocation for navigation, the Old World fruit bats have large eyes, navigate by sight, and locate ripening fruit by sight and smell. They have very good night vision because of the projections from their eyeballs, which penetrate the retina and provide a greater surface area for rods—the receptors for black-and-white vision. In keeping with their nocturnal habits most bats do not have cones in their retinas, and as night vision is not a major sense in the insectivorous bats they do not have the retinal projections. The fruit bats do not avoid sunlight, and many roost fully exposed to the sun, cloaked in their wings, occasionally waving one to create an air current and reduce body temperature; they are often seen in flight during daylight, moving between roosting places. However, the large fruit bats usually leave their roosts at twilight, and on very dark nights have blundered into objects difficult to negotiate such as telegraph wires, and even barbed wire fences.

Bats in Flight

Bats are the only mammals adapted for flight. Although several mammals can glide for short distances on fur-covered side flaps, bats are the only ones that can truly fly. Their wings are large membranes that extend from the sides of the body and hind legs, enclosing the elongated bones of the forelimbs and the phalanges of the hands, while leaving the first digit or thumb free. Their feet are also free. In most species there is also a membrane between the legs which encompasses the tail and is called the interfemoral membrane, whereas in the family appropriately called the free-tailed bats the tail extends beyond this membrane. Another membrane— antibrachial—links the neck with the forearm, and inside the ankle joint a cartilaginous spur called the calcar spreads the tail membrane. The framework of their wings is formed by the typical bones of the vertebrate forelimb modified to such a degree that the third finger is very elongated and can equal the length of the bat's head and body. The wings are very thin and elastic and are simply two layers of skin enclosing nerves and blood vessels. Their toes are well clawed to support their weight when roosting; their thumbs also have strong claws, and in some species of *Pteropodidae* the second digit is also clawed. Bats' forearms can only act as wings because of several structural skeletal changes that provide rigid support. The shoulder girdle is held firm by an enlarged collarbone, the vertebral column is fused, and the ribs are flattened. To power the wings the sternum is ridged, a similar arrangement to the flying bird's keeled breastbone, as a base for the powerful muscles that pull the wing downward, but unlike birds the flight of bats involves both the wings and legs working in unison. The bat's large thorax holds an enlarged heart and powerful lungs which provide the endurance needed for hours of nonstop flight.

Bats at Rest

All bats rest by day. Most species hide, seeking shelter where the microclimate and darkness provide both security and a suitable environment, such as caves, mine shafts and tunnels, hollow trees, under loose tree bark, and in the attics and lofts of buildings, especially church towers. Caves are certainly the bat's favorite sleeping places, and in the north they also serve as hibernation roosts, as they offer protection from inclement weather and provide constant temperature and humidity levels irrespective of the external conditions. Bats need a humid environment, so a damp, cool cave is ideal for overnight or overwinter. Their wing membranes become dehydrated if the humidity drops below 85 degrees and if they do not have access to water. Others roost in the open, some hanging partially hidden beneath leaves, others fully exposed to the sun and predators in communal tree roosts. Most bats

Hemprich's Bat *An insectivorous Palaearctic species, showing the typical bones of a vertebrate forelimb modified for flight, with extremely elongated fingers, all enclosed together with nerves and blood vessels between two layers of skin. The tail in this species is also enclosed—by the interfemoral membrane.*
Photo: Courtesy Ivan Kuzmin

hang by their feet, head downward, but some cling on with all four limbs and some may rest on ledges.

Bats do not have many predators, but it is surprising that their most serious predation occurs not when they are roosting and are seemingly vulnerable, but during flight—although mainly when they leave their caves en masse. Their regular predators, which specialize in catching bats, are birds like the bat falcon (*Falco rufigularis*) and the bat hawk (*Macheiramphus alcinus*), and snakes such as the Cuban boa (*Epicrates angulifer*) and Borneo's marbled rat snake (*Elaphe taeniura ridleyi*), which seize them as they enter or exit their roosting caves. Within the caves the major threat is to those that fall to the cave floor, usually the young or ailing individuals, where they are at the mercy of cockroaches and other aggressive invertebrates, plus rats and civets.

The smaller tropical bats that roost individually in the open in trees, in birds' nests, or in hollow tree trunks are certainly at risk from diurnal snakes and arboreal mammalian predators such as the tayra in South America, genets in Africa, and the linsangs in Asia. The larger fruit bats, which roost openly in exposed colonies, may suffer predation by hawks and perhaps snakes, but it is believed their unpleasant odor may deter some predators; yet they are still eaten by humans in the South Pacific. The fruit bats or flying foxes wrap their wing membranes around their bodies, shielding themselves and their young from the direct sun and fanning their wings to lower their body temperature. They are also often seen flying between roosts in bright sunlight. They are the exceptions among bats, however, for most

African Leopard *Leopards are very good climbers. They rest in trees during the day and also climb after dark to seize roosting baboons or panic them into falling, when they can easily catch them as their senses are far more attuned to night activity.*
Photo: Photos.com

Tomato Frog   *A rare frog from northern Madagascar with powerful skin gland toxins to deter predators. This is a female; the males are a dull yellowish-orange.*
Photo: Harcourt Index

Painted Reed Frog   *These small frogs are from southern Africa, where their populations are variable in color and patterning. They are nocturnal but frequently sunbathe.*
Photo: Harcourt Index

Banded Bullfrog   A chubby ant-eating frog from Southeast Asia, the banded bullfrog appears at night after heavy rains, when the males inflate themselves and float on the water's surface while calling to attract a mate.

Photo: Loong Kok Wei, Shutterstock.com

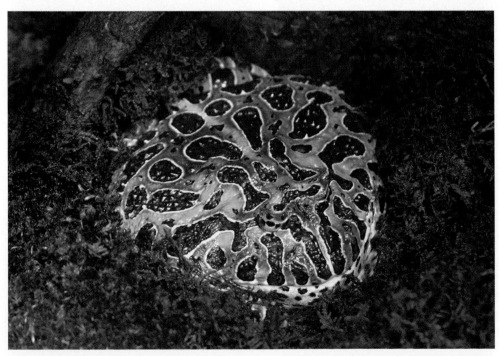

Ornate Horned Frog   The most colorful large amphibian, the ornate horned frog is so terrestrial it may drown in deep water. A sedentary species, it waits in ambush for its prey to pass close enough to seize.

Photo: Steffen Foerster, Shutterstock.com

Green Tree Python *The Green tree python is a nonvenomous constricting tree snake from New Guinea and Australia's Cape York Peninsula. Here the heat-sensing pits along the lower jawline can be clearly seen.*
Photo: Brad Thompson, Shutterstock.com

Young Green Tree Python *Young green tree pythons are brick-red or lemon-yellow and do not attain their adult green base color until they are about eight months old.*
Photo: Brad Thompson, Shutterstock.com

Tokay Gecko *One of the largest geckos and a very popular pet species, the tokay gecko is also one of the most colorful of all the nocturnal species. A native of forested regions of southeastern Asia, it depends upon its highly developed vision and acute hearing to acquire its insect prey after darkness falls. Unlike other lizards, the geckos have vocal chords and are very noisy.*
Photo: Courtesy Richard Ling

Matschie's Tree Kangaroo   Climbing down a tree is a laborious business for tree kangaroos, which are otherwise good climbers and great jumpers. This species, the most colorful of all the kangaroos, lives in the mountain forests of New Guinea's Huon Peninsula.

Photo: Clive Roots

Tamandua   The mainly nocturnal tamandua is one of two tree anteaters in the neotropical forests, although it actually eats more termites than ants. It has a prehensile tail, a typical anteater's long snout, and strong forelimbs and claws for breaking termitaria apart.

Photo: Tony Hathaway, Dreamstime.com

Brush-tailed Possum   A common Australian marsupial that was introduced into New Zealand in 1837 to increase the supply of pelts for the fur trade, the brush-tailed possum has been very damaging to the forests and native wildlife. The current population of about 70 million eat an estimated 7 million tons of vegetation annually.

Photo: Stuart Elflett, Shutterstock.com

Red Panda *Although it lives in the forests of western China and eats bamboo, this panda is a relative of the raccoon, not the giant panda, from which it also differs in its nocturnal tree-climbing behavior.*

Photo: Lynsey Allen, Shutterstock.com

Red Fox *The senses of smell, hearing, and vision are highly developed in this wild dog, which has a wide distribution across Europe, Asia, and North America, plus North Africa. It is normally nocturnal.*

Photo: Wesley Aston, Shutterstock.com

Axis Deer   A common animal in India and Sri Lanka, the axis deer has been introduced into several countries and is now plentiful in Texas. Although diurnal, it is preyed upon at night in its native lands by the nocturnal Bengal tiger, leopard, and the pack-hunting red dogs. Striped hyenas and golden jackals often take its fawns.

Photo: Steffen Foerster, Shutterstock.com

Springboks   A southwest African arid-land gazelle, the springbok is active by day but is vulnerable at night to lions, leopards, and both spotted and brown hyenas. Jackals, ratels, and the caracal lynx take its calves.

Photo: Photos.com

East African Prey   Here are three of the African grassland's major prey animals—the impala, the zebra, and the giraffe—whose calves are especially vulnerable to large predators. Although totally diurnal, these ungulates are fair game at night to lions, leopards, and spotted hyenas.

Photo: Mike Carlson, Dreamstime.com

African Lion   The African lion is Africa's most impressive carnivore and the only social species of wild cat and communal hunter. With superb vision, hearing, and sense of smell, plus tremendous jaw power, lions are the supreme night hunters. The daylight hours are spent resting or socializing.

Photo: Steffen Foerster, Dreamstime.com

seek seclusion when they are sleeping during the day, and while doing so experience a considerable drop in their body temperature, often to the point of semi-torpidity. This state of partial suspension of their sensibilities—lowered heart rate, lowered temperature, reduced respiration—is very similar to long-term winter hibernation, and the purpose of both is to save energy.

Most bats have somber coloration and are difficult to observe on a cave wall or inside a hollow tree trunk. Even the Asian painted bat (*Kerivoula intermedia*), the most colorful species—a lovely scarlet or bright-orange bat with black wings—is difficult to see in the tropical forest foliage of Southeast Asia. Solitary roosters are rare, and numerous species that were once believed to roost on their own have since been seen in company of others. In some species males roost solitarily away from the females. Known solitary roosters include Keens little brown bat (*Myotis keenii*) of North America and the African painted bat (*Kerivoula argentata*), which roosts in birds' nests. Other bats have made equally unusual choices for their daytime roosting sites. The Asian bamboo bats (*Tylonycteris*) sleep inside bamboo stems, their flattened heads allowing them to squeeze through cracks in the canes, and they have suction pads on their feet to cling to the inside of the bamboo. Other species of African painted bats sleep in curled plantain fronds, a habit now adopted by pipistrelle bats in banana plantations. The fulvus leaf-nosed bat (*Hipposideros fulvus*) of India and Sri Lanka has been found sleeping in crested porcupine burrows, and the tent-building bat (*Uroderma bilobatum*) and several species of *Artibeus* bats bite partly through a palm frond so that the end hangs down, making a "tent," which they may use for several weeks.

Some of the Species

Frugivorous Bats

In the tropical New World there are many fruit-eating or frugivorous bats. They are members of the family *Phyllostomidae*, which also contains other species with insectivorous and carnivorous habits. The fruit-eaters include the Cuban fig-eating bat (*Phyllops falcatus*) and the Jamaican fruit-eating bat (*Artibeus jamaicensis*), and there are also aberrant species in which the tongue and snout are adapted for a diet of nectar and pollen, and they can hover like hummingbirds in front of flowers.

The fruit-eating bats living in the warmer regions of the Old World belong to the large family *Pteropodidae*, which contains the largest bats. Fruit forms the bulk of their diet but their relatively small body size does not allow them to waste time digesting fiber, and to meet their energy requirements they swallow only soft fruit pulp and fruit juices. After crushing the fruit between their tongue and ridged palate, they spit out the indigestible seeds, pith, and skin. They also chew flowers to extract the nectar and then eject the fiber. They may hang onto an adjacent branch to eat, cling directly onto the fruit, or carry it off to eat in flight or at their roost, and are therefore important seed dispersal agents. Such a diet is easily digested and food passes through their gut in thirty minutes. There are aberrant species in this group also, which feed almost exclusively on pollen and nectar.

Flying Fox   *A highly colonial species, the flying foxes are nocturnal like all bats, but they also have good vision and search for ripe fruit at night by sight and possibly smell. Unlike most bats, during the day they hang from trees in the full sun with their large wings wrapped around them, flapping regularly to keep cool.*
Photo: Nicholas Rjabow, Shutterstock.com

Indian Flying Fox (*Pteropus giganteus*)

The Indian flying fox is one of the largest members of the family *Pteropodidae*, which has over 170 species. It is a most impressive bat, dark-brown in color with a patch of grayish-yellow between the shoulders, and has a wingspan of 48 inches (1.2 m), a head and body length of 16 inches (40 cm), and weighs up to 3 pounds (1.3 kg). It can be seen throughout the Indian subcontinent south of the Himalayas, plus in Myanmar and Sri Lanka. Flying foxes have large eyes with a high concentration of rods for black-and-white vision, and are most dependent upon sight and smell for locating their food. They are the most obvious bats in the tropics because of their habit of roosting colonially in trees, fully exposed to the weather and unfortunately to human hunters, as they are eaten in some countries. Their diet is entirely fruit and blossoms, wild figs being a favored food, but they also cause damage in orchards and plantations of cultivated fruit. Their day roosts or "camps" may contain thousands of bats, which kill the undergrowth beneath their favorite trees, and eventually the trees themselves when the soil is enriched with their copious droppings.

Straw-colored Fruit Bat (*Eidolon helvum*)

A single species in its genus, the straw-colored fruit bat is a smaller animal than the flying fox but, like it, is a very social creature. A native of Africa south of the Sahara, Yemen, and Madagascar, it is yellowish-brown with a brownish-orange belly and a bright-orange neck, and the males are more brightly colored than the females. It has a head and body length of about 7 inches (18 cm) and weighs 10 ounces (283 g); it has long and narrow wings spanning 26 inches (66 cm) that are an adaptation for long-distance flying, which it does in its constant search for flowering or fruiting trees. It is a serious pest in date and fruit plantations. An animal of the forest and open woodland, it apparently locates its food by vision, and roosts in large colonies in trees or in caves. In Zambia's Kasanka National Park a reputed 5 million bats roost in a small area of forest, where they attract many predators. Eagles and hawks pluck them from the trees, and crocodiles and monitor lizards wait beneath for those that get knocked down in their constant squabbling and jostling for the best roosting position.

Seba's Short-tailed Bat (*Carollia perspicillata*)

This is a small, dark-brown bat from Central America and northern South America, and several islands in the Caribbean's Lesser Antilles, where it lives in lowland evergreen forests. *Carollia* has a short wingspan and weighs only $\frac{2}{3}$ ounce (20 g) when adult; it is primarily a fruit- and nectar-eater but also eats insects, and when food is short it enters a state of torpor with lowered body temperature, despite living in the tropics. It forages close to the ground, and has an excellent sense of smell for locating ripe fruit and flowering plants. The fruit of the piper tree, a member of the pepper family, is a favorite food and, to illustrate their keen sense of smell, many have been caught in mist nets impregnated with oil from the piper fruit. Seba's bat seeks a darkened place for roosting, preferring hollow trees, caves, mines, and buildings, where it gathers in groups of up to 100 individuals. Although colonial, it is a very territorial bat, with males dominating a number of females and defending their harems against other males.

Insectivorous Bats

The insect-eating bats are the most plentiful in terms of species and numbers, and they occur throughout the world, excluding the polar regions. All bats living in northern temperate environments lack the constant supplies of fruit and nectar, amphibians, and fish available to the tropical species, so they are by nature insectivorous, and the disappearance of insects with the onset of winter forces them to hibernate or migrate. Despite the availability of so many other forms of food in the warmer regions of the world, however, many species of bats living there are also insectivorous.

The largest family of insectivorous and carnivorous bats is *Vespertilionidae*, containing about 350 species, including several of the most frequently seen bats in the Northern Hemisphere, such as the little brown bats (*Myotis*) and the pipistrelles (*Pipistrellus*), which hibernate for several months each winter in the cooler parts of their range. They roost daily in caves, mine shafts, and buildings, seeking a dark and secure place as they are very light-sensitive, and normally hang onto a vertical surface. The range of a few species extends into the northern forests, and in North America both the hoary bat (*Lasiurus cinereus*) and the red bat (*Lasiurus borealis*) spend the summer months as far north as the northwestern shores of Hudson's Bay, while in Eurasia the northern big brown bat (*Eptesicus nilssoni*) ranges into the Siberian coniferous forests. These bats roost on tree trunks during the day.

Bats vary in their method of insect-catching, although most species are "hawkers," which catch insects in mid-air. They seldom catch them in their mouths, however, but scoop them up with their large wing membranes and smaller tail membranes, and then bring them to within reach of their jaws. The lesser horseshoe bat (*Rhinolophus hipposideros*) flies close to the ground and gleans insects from stones and branches with its mouth. Others are ambushers, which seek their prey while hanging upside down, their flexible hip joints allowing them to turn 360 degrees, and when their returning echoes reveal an approaching insect they swoop out to seize it. A bat's feeding habits are usually revealed by the structure of its wings. High fliers like the sheath-tailed bats (*Coleura*), which seek fast-flying insects, have long, tapering swallow-shaped wings, whereas the horseshoe bat and other low and slow fliers have rounded, almost stubby wings.

Little Brown Bat (*Myotis lucifugus*)

This is the common bat of North America, ranging from southern Canada through the United States into northern Mexico. Its coloration varies, but always in shades of glossy brown, with paler underparts, and it has a wingspan of 10 inches (25 cm), small black ears, and a short tragus—the cartilaginous flap in front of the ear opening, which improves the receipt of sound. Northern populations of this species either hibernate where they have spent their summers or migrate farther south and then hibernate for a shorter period, returning to their regular roosts in spring. In summer they roost in attics, beneath house siding, and under bridges in groups, but single animals, usually males, hide under bark and in rock crevices. They hibernate for the winter months in caves where the humidity is high and the temperature remains constant a few degrees above freezing. Little brown bats are totally insectivorous, and prefer to hunt over wet areas, where insects are usually more plentiful. They have an unusual method of delayed reproduction, similar to delayed implantation. After mating in the fall, and sometimes while in winter hibernation, the sperm remains inactive within the female until spring, when fertilization occurs and the single baby is born after a gestation period of fifty-five days.

Common Pipistrelle (*P. pipistrellus*)

This is the smallest European bat, once a common species in Europe and England, but now in rapid decline due to habitat loss and the effects of toxic chemicals. It has a dark reddish-brown back with yellowish-brown underparts, has a wingspan of 9 inches (23 cm), and weighs about $\frac{1}{7}$ ounce (4 g). It lives virtually everywhere, in open woodland, parkland, farmland, marshes, and in towns. An early riser, the first bat seen in the dusk of early evening, it is less sensitive to daylight and has been seen flying during the day. A fast and agile flyer, the pipistrelle hunts moths, gnats, and other flying insects, generally no more than 30 feet (9 m) above ground. After feeding for a few hours it usually goes back to its roost to rest and then appears again later, consuming up to 3,000 insects in the course of a night. From May to September the females gather in maternity roosts, with an average of 150 bats per roost, and the males remain solitary during this period. The young, usually twins, can fly when they are four weeks old and are weaned two weeks later, at which time the colonies disperse. From mid-October to early April these bats hibernate in buildings, attics, church belfries, and cellars.

Mexican Free-tail Bat (*Tadarida brasiliensis*)

Despite its name the Mexican free-tail bat is the most common member of the family *Molossidae* in the southern United States, and has an even larger range that also includes all of neotropical America down to Argentina and the West Indies. It has velvety chocolate-brown fur, with a wingspan of 13 inches (33 cm), and true to its name its tail extends well beyond the edge of the interfemoral membrane. It has a wide range of habitat—from desert to forest and from sea level to 9,000 feet (2,745 m). The free-tail bat is a fast flyer that roosts in huge colonies—the largest concentrations of mammals in the world—the most famous being those in New Mexico's Carlsbad Caverns, whose flights are so dense they can be tracked on radar. It is estimated that 100 million bats live in Texas, eating 1,000 tons of insects nightly, and as many of these are agricutural pests such as cotton boll moths and cut-worm moths, the bats are considered highly beneficial animals. Large colonies roost in caves, mines, and under bridges, whereas smaller groups generally sleep in buildings and trees. It has a short period of inactivity during old weather and some individuals migrate to Mexico for the winter.

Carnivorous Bats

Many of the bats in the suborder *Microchiroptera* have diverged from the basic insect-eating habits of the group to an omnivorous diet and even to carnivory. The carnivorous bats eat a wide variety of vertebrates, including fish, amphibians, nocturnal reptiles, birds, and small mammals, whereas the omnivorous species eat many of those prey items also, but add fruit and nectar to their diet. There are generalized carnivores such as the false vampire bats, which are not the blood-letting vampires, and whose attack is terminal as they eat their whole prey rather than just drink its blood. There are also some highly specialized carnivorous species, such as the frog-eating bat (*Trachops cirrhosus*) of tropical America, which plucks tree frogs off branches while in flight; the fisherman bat (*Noctilio leporinus*); and of course the vampire bat (*Desmodus rotundus*), which lives solely on blood, and whose attack is not terminal unless it hosts rabies. The habits of some are hardly bat-like, such as the ambushing African false vampire bat (*Cardiodema cor*), which hangs from low branches in the forest and drops down onto its prey on the ground.

Linnaeus's False Vampire Bat (*Vampyrum spectrum*)

This is the largest bat in the New World, also known as the spectral bat, which occurs throughout Central and South America. It has a wingspan of 36 inches (92 cm), weighs 6.5 ounces (185 g), and is reddish-brown with a paler belly. It has very large ears, an elongated snout, and a pointed nose leaf which curls forward. It is a bat of the forests—deciduous, evergreen, and secondary-growth woodland—usually in moist areas. An aggressive and versatile predator, it hunts mainly in the evening, seeking birds, other bats, and small mammals such as mice and mouse opossums, which it drops onto from a branch or the air, seizes its victim by the head, and sinks its long and sharp canine teeth into its brain. It can eat its prey completely as it also has strong molar teeth for crushing bones. The false vampire bat lives in tree hollows in family groups, comprising the adult pair and several young, and when they leave the roost to hunt, an older youngster stays behind to babysit the latest offspring, and the hunters bring back food for them.

Common Vampire Bat (*Desmodus rotundus*)

The vampire bat is the most notorious of all bats, a blood-drinker or sanguivorous animal and a vector of the rabies virus, whose activities cause human deaths and heavy losses of domestic animals, especially cattle. An aberrant member of the family *Phyllostomidae*, which includes several small fruit- and nectar-eaters, it occurs throughout the neotropics from Mexico to Argentina, and on Trinidad. The only species in its genus, it is a small bat with a head and body length of 3 inches (8 cm), a wingspan of 12 inches (30 cm), and weighs just 1.2 ounces (36 g). It is grayish-brown above with paler underparts and has long, pointed ears. The vampire bat roosts both solitarily and colonially, in caves, hollow trees, and abandoned buildings, and flies low on its blood-hunting trips, emitting low "whispering" calls that have just a fraction of the sound energy of the insectivorous species.

In keeping with their liquid diet these bats have a short esophagus and a narrow stomach, and their front teeth are adapted for scooping out flesh, being slightly recurved and having sharp cutting edges. They seek areas of bare skin or sparsely furred or feathered regions of their prey, and when a suitable target is

Vampire Bat *The blood-drinking vampire bat lands near its intended victim and then scuttles across the ground and searches for an area of exposed flesh. Scooping out a piece with its sharp teeth, it then laps and sucks the blood, an anticoagulant in its saliva ensuring the continued flow. In the process it may transmit paralytic rabies to its victim.*
Photo: Michael Lynch, Shutterstock.com

discovered they alight nearby and scuttle silently and "crab-like" over to it, and without waking their victim, they scoop out a tiny piece of flesh and then drink the flowing blood. An anticoagulant in their saliva ensures the blood flows freely, and it usually continues to do so for some time after the bat has finished its meal. The blood is consumed by a combination of lapping and sucking, with the down-curved edges of the tongue forming an open-bottomed tube which is extended and retracted slowly. A vampire bat can drink up to 20 ml of blood nightly, but may gorge so much that it is unable to fly, although it can still jump and run with surprising agility.

Wild animal victims include deer, tapir, large roosting birds, and even snakes. Domestic animals, such as horses, cattle, and poultry, are attacked, but sleeping dogs rarely fall victim as their sensitivity to high-frequency sounds warns them of the vampire's approach.

Fisherman Bat (*Noctilio leporinus*)

One of only two members of the family *Noctilionidae*, and a highly carnivorous species, the fisherman bat is another very specialized neotropical animal, occurring from Mexico south to Brazil and the West Indies. It is a medium-sized bat, with a head and body length of 5 inches (12.5 cm) and maximum weight of 2.6 ounces (72 g), and has long and narrow wings that span 20 inches (50 cm). A colonial species, it sleeps in small groups in caves (including sea caves), hollow trees, and buildings. The fisherman bat is sexually dimorphic, the males having orange-rufus upperparts and the females grayish-brown, with pale underparts in both sexes. *Noctilio* has long hind legs, large feet and claws, and fishes over both freshwater lakes and rivers and the sea, skimming low and trailing its legs just above the surface. It flies with its head pointed downward to the water and it was always assumed that echolocation was involved in fish location and capture, but only of fish that had broken the surface, as the supersonic calls could barely penetrate water. Now it is thought that the hanging lower lip directs the sound waves onto the water's surface and that echoes are possibly returned by the swim bladders[3] of fish just below the surface. When successful it either eats in flight or takes its catch to a roost. In addition to fish it also eats frogs, crustaceans, and insects.

Notes

1. Variation in the frequency of vibrations.
2. A Hertz is a unit of frequency that equals one second. It was formerly called cycles per second (cps).
3. The swim bladder or gas bladder is an internal organ that helps fish maintain their buoyancy—to stay where they are, to descend or to ascend—without swiming.

7   The Insect-Eaters and Their Allies

Insects are the most abundant source of usable[1] animal life on land in terms of both variety and numbers, and most of the early mammals that began evolving in the Mesozoic Era were not only nocturnal but were also primarily insectivorous, taking advantage of the rich food resource the insects provided. Some mammals still rely completely upon insects, but surprisingly they are not members of the order *Insectivora* (the insect-eating mammals). In fact, the three nocturnal groups of that order included here, the hedgehogs, tenrecs, and shrews, have evolved more liberal tastes and are virtually carnivorous, and could manage quite well without insects. In contrast there are other mammals, which are not members of the order *Insectivora*, that are purely insect-eaters.

The true insect-eating mammals are the bats, which are included in the previous chapter, and the anteaters—which would starve to death if denied food small enough to enter their tiny mouths. They have evolved to survive on a diet composed solely of ants and termites, mainly the latter, and the names of most reflect their food preferences. They belong to the orders *Xenarthra* (the anteaters), *Pholidota* (the pangolins or scaly anteaters), and *Tubulidentata* (the aardvark, also called the ant bear), and all except the aardvark are toothless.

Although birds are usually considered the most efficient controls of the insect hordes, for sheer quantity of individual intake none can compare with these purely insectivorous mammals, which can consume a gallon of ants or termites daily. The bats, of course, the subject of the previous chapter, are the greatest eaters of insects. A few anteaters live in trees, but most are terrestrial animals that seek their food on the ground. Vision is not an important sense for them, but their sense of smell is highly developed.

The nocturnal allies of these pure insect-eaters are a very mixed group, and include three other families of the *Insectivora*, namely, the hedgehogs (*Erinaceidae*), the tenrecs (*Tenrecidae*), and the shrews (*Soricideae*), all of which have evolved jaws and teeth able to cope with a wider range of foods, even ones that require biting and chewing, and they are carnivorous rather than insectivorous. Their diet includes other invertebrates (snails, slugs, worms) plus birds' eggs and nestlings, small lizards, and the nestbound young of mice and voles. In addition to these carnivores there are the armadillos and sloths, surprisingly now considered members of the order *Xenarthra* along with the "true" anteaters. The armadillos are terrestrial and have similar carnivorous habits to the shrews and hedgehogs, but the sloths are arboreal and have become adapted for a totally vegetarian diet.

■ THE INSECT–EATERS

Some of the Species

Anteaters

These animals eat only ants and termites as they are restricted to a diet of foods that can be drawn into their narrow mouths on a long and sticky protractile tongue, and they are all toothless. Even though they evolved on distant continents they resemble each other in the form and behavior associated with gathering these insects, their similarities arising as a result of a common lifestyle, not any ancient relationship. Unlike the fairly recent behavioral and anatomical changes that animals like the sloth bear and aardwolf have evolved as a result of their liking for ants, the anteaters' extreme modifications for gathering and digesting ants have been developing for a long time, and fossils of their ancestors date from the Eocene Epoch about 50 million years ago.

Despite their difference in size, the four species of anteaters, of which three are nocturnal, have typical anteater characteristics—a long snout, a very long tongue, powerful forelimbs and claws for opening the hardened, concrete-like casing of termite mounds, and thick skins to protect them from their prey. They have a very good sense of smell to locate their food, but their small ears and tiny eyes indicate that hearing and sight are poorly developed. To keep the long tongue well lubricated they have enlarged salivary glands, plus adaptations of the tongue and stomach. Their tongues have tiny, backward-pointing papillae to aid ant collection and their stomachs are well adapted to cope with hard-shelled insects, for they have muscular walls lined with horny epithelium instead of the usual mammalian mucus membrane. Small stones, perhaps swallowed intentionally, assist in grinding the insects in the manner of a bird's gizzard. It is also possible that these mammals produce the enzyme chitinase, which hydrolyzes the invertebrate exoskeleton, which is made of chitin. In most locations their specialized diet ensures they have few, if any, natural competitors.

Anteaters are said to be myrmecophagus—Greek for ant-eating—as their diet comprises solely insects of the orders *Hymenoptera* (ants) and *Isoptera* (termites), actually mainly the latter, which are often confused with ants. For such relatively large mammals to survive on a diet of tiny insects requires a very efficient gathering system. Once the hard outer shell of the termitarium has been broken away, the anteater's long and sticky tongue penetrates deep into the passageways collecting all the insects that come into contact, together with considerable debris. Anteaters also feed on ant columns on the march. It has been estimated that the tamandua eats about 9,000 ants and termites daily,[2] and although no figures are available for the aardvark, it must find sufficient food to support a body weight ten times greater than the tamandua.

The anteaters are animals of the New World tropics, and for many years were called edentates, of the order *Edentata*, but they are now considered members of the order *Xenarthra*.

Silky Anteater (*Cyclopes didactylus*)

This is the smallest and most arboreal of the neotropical anteaters, a tiny animal just 6 inches (15 cm) long with a tail measuring 8 inches (20 cm), and a weight of ½ pound (227 g). It has soft and silky fur of a beautiful burnished gold, which is actually very cryptic when the anteater is curled asleep on a branch in the sun-dappled foliage. It is often seen in the silk cotton or kapok tree (*Ceiba pentandra*), which has large seed pods with silky floss that serve as camouflage, and it rarely comes down to the ground. Its main predators are harpy eagles and crested eagles during daylight and the spectacled owl at night. The silky anteater is an inoffensive animal, but it can inflict deep cuts with its sharp-clawed forefeet, which it raises in front of its face in a praying position and then jabs down swiftly. It has a prehensile tail—nature's finest tree-climbing aid—and extra-long soles, which not only give a good grip on branches but are jointed and thus give a degree of opposability and an even better grip so it is difficult to dislodge. *Cyclopes* gives birth to a single baby annually, and may leave it in a tree hole at night when it is foraging, but the father is also reported to carry the baby.

Tamandua or Lesser Anteater (*Tamandua tetradactylus*)

The tamandua is a medium-sized anteater, with a head and body length of 24 inches (61 cm), a 20-inch-(50 cm)-long tail and weighs 11 pounds (5 kg). It has short and coarse fur, which is longer on the back and tail, and is usually yellowish-brown with a black "vest" or "saddle" on animals from Central America and the southern parts of their range in South America. But individuals show considerable color variation, with both melanistic and all-blonde mutations. Its tail is prehensile, and is furred on top only at its base, the rest being naked and blotched black and white. Completely toothless like the other anteaters, the tamandua depends upon the muscular, gizzard-like section of its stomach to grind the chitinous prey and assist the acids in the digestive process. It has strong forearms and long and powerful claws, and to prevent forcing the large claws back into the footpads it walks on the sides of its hands, unlike the diurnal giant anteater, which walks on its knuckles. The forelimbs and claws are not its only defense against predators, as it can discharge a repulsive odor when alarmed. Although mainly an arboreal animal of the tropical forests that sleeps in a tree hole during the day, the tamandua also lives in grasslands, where it sleeps on the ground. It is mainly nocturnal but is occasionally active in daylight (see color insert). In Mexico the very similar Mexican tamandua (*T. mexicana*) may be a race of *T. tetradactylus* rather than a separate species.

Pangolins

The pangolins are ant-eating mammals that were once taxonomically linked with the other toothless species in the old order *Edentata* (now *Xenarthra*) but now occupy their own order, *Pholidota*, as it is believed that any similarities between them resulted from a common way of life, rather than ancestral relationships. They are unmistakable animals, with long and tapering bodies totally covered above with hard, sharp-edged scales, making them appear more reptilian than mammalian. The scales, which are composed of cornified epidermis that grows continually to replace wear and tear, overlap each other in adult animals and can be raised and lowered. Their bellies, the insides of their legs, and sides of the face are scale-less. Pangolins walk on their knuckles with the claws curled beneath to protect them, and they are at home in the forest, savannah, and scrub—wherever ants and termites can be found. They live in sub-Saharan Africa, and thoughout Asia from India to China and south into Southeast Asia. They vary in their activity, most of the seven species being nocturnal, with the others being equally active during daylight and at night.

Ants and termites are gathered by the pangolin's long and sticky tongue, which in the larger species may be 20 inches (50 cm) long, and is contained within the animal's body in the chest cavity and anchored to the pelvis. The tongue is lubricated by extremely large salivary glands which extend into the neck almost to the animal's shoulders. Pangolins are completely toothless, and their very large stomachs with thick, muscular walls act like a bird's gizzard to grind the hard-bodied insects. They have small eyes and sight is not well developed; they rely more on their sense of smell to find food and mates. Their hearing is apparently quite good, although the four African species have no pinna or outer ear to collect sounds, and in the three Asiatic species there is just a ridge of hard skin around the external ear.

African or White-bellied Pangolin (*Manis tricuspis*)

This is the long-tailed, arboreal pangolin of central Africa, ranging throughout the tropical rain forest from Senegal to Kenya and south to Zambia. It is the smallest of the pangolins, weighing about 4 pounds (1.8 kg), with a head and body length of 18 inches (46 cm) and a slightly longer tail, which is prehensile in keeping with the animal's tree-living habits. It has small eyes and thick eyelids, and its brown overlapping scales are three-pointed, hence its specific name *tricuspis*. Totally nocturnal, the African pangolin shelters in hollow trees by day, and when out at night it occasionally comes down to the ground, where it walks on all fours, or on its hindlimbs with forelimbs hanging free. It is now a rare animal, due mainly to forest clearance and the loss of its habitat and food source; it is considered a delicacy by the native people—a favorite form of "bush meat." It is also sought for medicinal purposes and in West Africa for juju[3] concoctions, and there is a huge international trade in pangolin skins. Its major natural predators are leopards and pythons.

Indian Pangolin (*M. crassicaudata*)

The Indian pangolin is a terrestrial species from Pakistan, India, Bangladesh, and Sri Lanka, where it is still fairly common only in protected areas, but has been overhunted elsewhere. It is one of the larger species, in which adult males (which are larger than the females) have a head and body length of 28 inches (71 cm), and a long tail which measures 18 inches (46 cm); it is surprisingly highly prehensile, which is unusual for a ground-dwelling animal. It has the typical powerful fore-limbs and very large foreclaws of the anteaters for ripping open termitaries. The Indian pangolin has dark yellowish-brown scales with sparse, bristle-like hairs between its plates. It is nocturnal and burrows down several feet to escape the heat of the day or takes shelter among rocks; it seems to have no habitat preferences, occurring in rain forest, dry thorn scrub, on the open grassy plains, and on moun-tain slopes. Its single baby is born in the burrow, and stays there for the first month, after which it rides on its mother's back.

Aardvark

The aardvark (*Orycteropus afer*) is the only living species in its order, a unique survivor of ancient animals that originally lived throughout the Old World. It is an animal of the savannah, wooded grasslands, and bushveldt of sub-Saharan Africa, and its name is an Africaans word meaning earth pig, although it is also commonly called ant bear. The aardvark certainly resembles a pig more than a bear, with a heavy body, short thick legs, and a long head and snout, which is blunt at the end with pig-like nostrils. It has long, tubular ears and is sparsely haired, with dull brownish or yellow-gray hair through which its thick, pinkish-gray skin can be seen; but there the external resemblance ends, for its tail is long, naked, and very muscular, totally un-pig-like. The aardvark is the largest of the pure anteaters, its body and long tapering head measuring 4 feet 3 inches (1.3 m), and its tapering tail is 28 inches (71 cm) long. It stands 28 inches (71 cm) high at the shoulder and weighs 165 pounds (75 kg). The subspecies (*O.a. albicaudus*) from Damaraland is a very distinctive dark-brown animal with a white tail. The aardvark has four toes on its front feet and five on the hind feet; the toes are webbed and have strong, straight and blunt claws. It apparently prefers termites to ants and tears open termitaria; it also feeds on columns on the march, licking them up with its long, tapering tongue. Thoroughly nocturnal, it has small eyes and weak sight, which is compensated for by its well-developed hearing and sense of smell. Unlike the other anteaters, aardvarks have teeth—just premolars and molars—which grow continuously and lack enamel. These are hexagonal in shape and have a tubular pulp cavity, hence the name of their order, *Tubulidentata*, meaning tube-toothed animals. Aardvarks are powerful diggers, probably the fastest burrowers of all mammals, and they fold their ears back when tunnelling. They are nocturnal and live in burrows dug into the hardest soil, with short ones for resting and longer ones, up to 42 feet (12.8 m) in length, with several entrances and chambers for giving birth. Many wild animals, and people, have taken refuge from bush fires in aardvark burrows.

■ THE INSECT-EATERS' ALLIES

Some of the Species

Hedgehogs

Most insectivores adopted nocturnal habits as protection from predators, but some have since evolved an additional protection in the form of a body covering of short, sharp spines. Hedgehogs have achieved this, plus the ability to roll into a ball to protect their head, legs, and belly. They are animals of the Old World only, distributed across Eurasia and throughout Africa, living in wooded regions and desert, hibernating in the north for the winter and estivating in hot, arid regions to avoid high summer temperatures. All hedgehogs are nocturnal and hide during the day under piles of leaves and even in old termites' nests, or in holes that they dig with their powerful legs and strong claws. In the British Isles and Europe they live in wooded areas and farmland, generally in regions with rich soil and leaf litter, but in Africa and central Asia they are desert animals; they dig burrows to escape the heat of day and can survive without water for several weeks.

Hedgehogs are solitary and terrestrial, and although they are called insectivores, as they are members of the order *Insectivora*, their diet is decidedly carnivorous. They eat a wide range of small animal life including nestling birds, young mice and voles, worms and beetles, and are not averse to carrion. They are even cannibalistic; caged animals have eaten their dead companions and hedgehogs have been seen feeding on road-killed conspecifics. They are able to cope with such foods as they usually have between 36 and 44 teeth. Hedgehogs have poor eyesight, but their hearing is good, and most species have medium-sized, upstanding ears. They also have a well-developed sense of smell, which they use for finding food, for recognizing conspecifics, and for sensing danger. Their long snouts are highly tactile, and help to locate earthworms and other soil invertebrates.

European Hedgehog (*Erinaceus europaeus*)

A common animal in Europe and the Mediterranean islands of Sicily, Sardinia, and Corsica, this is the most familiar hedgehog, although unfortunately the most frequent sightings are of roadkills. When adult it is about 12 inches (30 cm) long

European Hedgehog *A nocturnal insectivore which is now carnivorous, the European hedgehog has expanded its diet from insects to include a wide range of small animal life. It sleeps during daylight in old rabbit holes or in leaf litter, protected by its coat of spines and the ability to roll into a ball.*
Photo: Dan Briajki, Shutterstock.com

and weighs just over 2 pounds (900 g). It has a tiny tail and short ears; its spines are dark brown with yellowish-white tips, are 1 inch (2.5 cm) long, and usually number about 5,000 per animal. It rolls into a tight ball to protect the head, legs, and belly when threatened, which is adequate defense against natural predators, but no deterrent to the people who have traditionally eaten them. European hedgehogs prefer deciduous woodland, farmland, and scrub, and although terrestrial can climb fences of wire mesh; during daylight they rest in old rabbit holes, among rocks, or under piles of leafy brush or leaf mould. They hibernate there also—for up to five months in the most northerly parts of their range. One of their most unusual habits has been called annointing, which involves rubbing toads against their spines and licking objects excessively until they frothed at the mouth, and then rubbing the froth onto their spines, similar behavior to "anting" in birds, which rub ants on their feathers. The reason for this behavior is unknown.

Long-eared Desert Hedgehog (*Hemiechinus auritus*)

A hedgehog of the grasslands and semideserts of Asia and North Africa, the desert hedgehog is a small species, about 10 inches (25 cm) long, but quite slim and weighing only 10 ounces (285 g). It has longer ears and longer legs than the other hedgehogs but its most distinguishing features are its white belly hair and its dense, dark quills, which stick out in all directions in a tangled mass. It lives in a region of very hot summers, and although its nocturnal and burrowing behavior and the ability to dispel body heat through its long naked ears provide some protection in midsummer, it may still estivate to escape intense heat. From November to March in the more northerly parts of its range it must then hibernate to avoid the low temperatures and lack of food, and it usually burrows beneath shrubs for its dormant period. In the southern deserts it does not become torpid although it may be less active than normal in winter. It is omnivorous and has a very varied diet of small animal and plant food, but water is scarce in its environment, and with little opportunity to drink it must derive most of its fluids from its prey. Captive specimens have survived for ten weeks without drinking.

Tenrecs

On Madagascar and its neighboring islands the tenrecs take the place of hedgehogs, and they are remarkably similar, even though they are not closely related. They are believed to be descended from small mammals that became established on the islands in the Cretaceous Period about 100 million years ago, just before Madagascar broke away from Africa. In the absence of competitors they evolved to fill the vacant niches, including those occupied by hedgehogs elsewhere and in doing so ended up resembling them, another case of convergent evolution; and like the hedgehogs some tenrecs can even role into a ball when threatened.

Tenrecs are currently divided into two subfamilies, the *Tenrecinae*, which contains the hedgehog-like spiny species; and the spineless and shrew-like members of the *Oryzoryctinae*. Where these animals are concerned taxonomy does have a close relationship with their captive husbandry, for the spiny species have been maintained and bred frequently in zoos and laboratories, whereas the shrew-like tenrecs have rarely been kept and are poorly known. The most interesting feature of the tenrecs is their fecundity, with the tail-less tenrec (*Tenrec ecaudatus*) being the most prolific of all mammals, producing up to thirty-two young per litter. They are terrestrial although they can climb, and most species are nocturnal.

The tenrecs are omnivores that use their snout to dig for food, and all store fat in order to hibernate during the Austral winter—from June to September—usually in burrows with the entrance plugged with soil. Tenrecs that were dug up in winter were cold to the touch, had very low breathing rates, and had neither food in their stomachs nor feces in the intestines. The body temperature of hibernating tenrecs is usually just 2°F (1.1°C) above the ambient temperature, but even when they are active they have variable body temperatures ranging from 75.2°F (24°C) to 95°F (35°C), a lot lower than that of other mammals, which average 98.6°F (37°C). They have poor sight, good senses of smell and touch, and their hearing is believed acute, at least in the common tenrec, which makes rapid tongue-clicking sounds that may possibly be a primitive form of echolocation for communication in the dark.

Tail-less Tenrec (*Tenrec ecaudatus*)

The tail-less tenrec occurs naturally in Madagascar, but has been introduced and established on other western Indian Ocean islands. It lives in a variety of habitat wherever undergrowth provides hiding places, but does not occur in Madagascar's arid south. It has a stout body with pointed, hedgehog-like muzzle, short limbs with the hindlimbs slightly longer than the forelimbs, a head and body length of 15 inches (38 cm), and weighs 4 pounds 4 ounces (2 kg). Despite its name it has a tiny tail, about ¾ inch (2 cm) long. It is grayish-brown or dark brown with a coat of dense hairs and spines, which are often hidden by long guard hairs that can be erected on the nape. This tenrec burrows to hibernate, with total torpidity—lowered temperature, heart rate, and respiration—and is one of this species that does not roll into a ball to protect itself. Long, tactile whiskers are probably this tenrec's most important sensory adaptation, and it also has sensitive hairs on its back, which detect vibrations. It has reasonably good eyesight, believed to be better at least than in the other species, and it scent-marks its territory to communicate with conspecifics, so it follows that it has a good sense of smell.

Lowland Streaked Tenrec (*Hemicentetes semispinosus*)

The lowland streaked tenrec is a tail-less species that reaches a length of 7½ inches (19 cm) and weighs up to 10 ounces (285 g). It is the most distinctive tenrec, with long and hairy spines, which are barbed and detachable; blackish-brown in color, with longitudinal chestnut stripes along the sides and back, and one from the nose to the back of the head. It has an erectable crest of spines on the nape. Predominantly an animal of the forests and secondary growth bushland, it needs deep leaf litter to support the earthworms that form the bulk of its diet, and which it locates with its tactile muzzle. Single tenrecs make short tunnels with a nest chamber at the end, but family groups make more sophisticated burrow systems, which may be several yards long. Mother and babies communicate by vibrating special quills on their backs, which is known as stridulation. Although terrestrial, the streaked tenrec is a good climber.

Shrews

Shrews are the most numerous of the insectivores with almost 300 species. They are tiny, mouse-like animals with beady eyes, tiny ears, and a very pointed muzzle; one of their members, the Etruscan shrew (*Suncus estruscus*), is the world's smallest mammal. They range throughout the Northern Hemisphere and also occur in the mountains of northwestern South America. The important criteria for shrew health is moist habitat and mild temperatures where the risk of dehydration is low, and although they do occur in regions with very harsh winters their high metabolism does not permit them to hibernate. Shrews are mainly nocturnal and hide under anything large enough to offer shelter during the day, but their constant need for food may also result in activity during the day. They are the most voracious insect-eaters in the northern coniferous forests.

Shrews have several unique characteristics. They are the only mammals with a venomous bite; they are undoubtedly the most highly strung of all mammals; and some use a form of echolocation like the bats. They are solitary creatures that have a very short lifespan (occasionally three years but usually less), but during that time they can give birth to several litters averaging seven young per litter. Their very nervous disposition and rapid pace of life is powered by a normal heart rate of about 800 per minute that can rise to well over 1,000 when they are severely stressed, which can cause their death. This frantic lifestyle needs continual sustenance, and they aggressively hunt a wide range of small animal prey, resembling the mustelids (mink and weasels) in their predatory habits rather than the other insectivores. They eat salamanders, small mice, nestling birds, worms, beetles and frogs, and the largest shrews can kill small rats. Normally, they eat their weight in food daily, but this rises to almost double their weight for pregnant females.

The shrews' small eyes are almost hidden by fur and they have very poor eyesight. Their senses of hearing and touch are well developed, but smell is their primary sense; they have dermal glands which emit a strong scent that is used for territorial marking and for attracting a mate. In addition to the sounds they make that are discernible to the human ear as tiny mouse-like squeaks, several shrews also emit supersonic calls that are reflected back from surrounding objects and from potential prey, and function as an echolocation system, although nowhere near as sophisticated as that of the insectivorous bats.

The shrews (and the duck-billed platypus) are the only venomous mammals, although only the platypus actually injects venom, through its spurs, thus meeting the true definition of venomous. The submaxilliary glands of some shrews secrete a toxic saliva that acts as a nerve poison, and must be "chewed" into the bite wound to function. It seems overkill to subdue an earthworm with venom, but several bites quickly incapacitate a worm which is then stored to eat later. They also store insects from which they have bitten off the legs to prevent their escape. Shrews do not hibernate, but remain active all year throughout their range, however low the temperature, making tunnels under the snow to continue their search for food. But they do become torpid during their resting periods, and their temperature drops to effect a savings of energy. Although some shrews are called musk shrews, all species actually have musk glands on their flanks, and the excretions from these are most obvious during the breeding season, when they can locate potential mates with their acute sense of smell.

Common or European Shrew (*Sorex araneus*)

A member of the long-tailed group of shrews, which are widespread and common across the Holarctic region, this species lives in Europe and northwestern Asia. It is mainly nocturnal but is also active during the day when food is scarce; it is active all year, making runways through the grass and burrowing into loose soil in summer and tunnelling beneath the snow in search of prey, even in the depths of winter. The common shrew has a long and slender body with a head and body length of about 3 inches (8 cm), a weight of just $\frac{1}{3}$ ounce (10 g), and a long flexible snout, tiny eyes, and small ears. Its mouse-like tail is about 2 inches (5 cm) long, and it has a silky, dark-brown coat with slightly paler underparts. The common shrew is a solitary and aggressive animal, and although terrestrial it can climb into low bushes. It is highly carnivorous and probes and sniffs to locate food in the soil, feeding voraciously on worms, insects, slugs, spiders, and carrion, and in turn is hunted by owls, weasels, stoats, and foxes. When disturbed the baby shrews follow their mother in line, each holding onto the tail of the one in front.

Short-tailed Shrew (*Blairina brevicauda*)

A common species from the eastern United States and southern Canada, the short-tailed shrew measures 3½ inches (9 cm) in length, but has a tail only 1 inch (2.5 cm) long. Adults weigh about ½ ounce (14 g), have a slate-gray coat with paler underparts, and have such tiny eyes that they are barely visible. It lives in a wide range of habitat, including forests, marshes, and grassland, and despite being considered a terrestrial animal is a good climber and has been seen in trees several feet above ground. The short-tailed shrew may be the only omnivorous species, for although it favors snails, beetles, and small mammals, it also eats nuts, berries, and seeds, and stores food for later use after immobilizing animate prey with its mild venom. A copulatory tie or "locking" occurs during mating in this species, caused by the unusual shape of the penis. The short-tailed shrew burrows more than the other species and spends much of its life underground in a system of tunnels, which it digs with its strong claws and cartilaginous snout, and it also uses the tunnels of other small mammals.

Armadillos

The armadillos, like the sloths, were members of the old order *Edentata* (the toothless ones) which was recently renamed *Xenarthra*, as they are far from toothless. They are all members of the family *Dasypodidae*, with a total of twenty species in several genera, including the three species of hairy armadillos (*Chaetophractus*), the six species of long-nosed armadillos (*Dasypus*), and the monotypic six-banded armadillo (*Euphractus*). Armadillos are restricted to the New World, and occur from the central United States to Argentina, where, at the north and south extremes of this range, they encounter cold winters and have evolved strategies such as fat deposition and the use of metabolic water for periods when food and water are scarce; but they have not yet learned to fully hibernate.

Their common name is derived from the Spanish for their armor-like body covering. They are certainly very well protected, at least from small predators, although they can be crunched by large cats, for in addition to their main body shell the outsides of their legs are covered with protective plates, their tail is completely encased, and they have a flat plate over the front of the head. These plates are composed mainly of horn, and are more akin to the shell of a tortoise than to any other mammalian body covering; they are also called scutes, like the tortoise's shell segments. They are connected by skin and are quite flexible, but in some species in the genus *Tolypuetes* flexibility is greatly increased by movable bands, which allows them to roll up tightly to protect their underparts. In several species hair grows through the gaps between the scutes.

Armadillos have a regular body temperature of about 91.4°F (33°C), much lower than the mammalian average of 98.6°F (37°C). They can burrow swiftly with their powerful legs and sturdy claws and are good swimmers. They have a keen sense of smell and reasonably good hearing, but their vision is poor and they have actually blundered into people who were standing still. Like the sloths they lack incisor and canine teeth, but have between fourteen and eighteen enamel-less cheek teeth in each jaw, which are open-rooted and continually growing.

Armadillos are purely terrestrial and occupy a wide range of habitat including grasslands, wooded savannah, and forest. There is also a great deal of variation in their activity patterns, and although they are mainly nocturnal and spend the day in their burrows, they may also appear during daylight; and they may be diurnal in cool weather and nocturnal during the hottest days of summer. Most species are either solitary or live in pairs, and when larger groups are seen it is usually because they have gathered to take advantage of a good source of food.

Although insects figure prominently in their diet, they also eat other invertebrates such as snails and worms, plus birds' eggs and nestlings, amphibians and snakes, and they eat ripe fruit. They are not quick enough to catch mice, but eat the babies when they find a nest, and carrion is very acceptable. They burrow beneath large carcases to find blowfly maggots, and the giant armadillo has been known to dig into fresh, human graves to feed on the corpses.

Nine-banded Armadillo (*Dasypus novemcinctus*)

This is the most common species, with the widest range of all the armadillos—from the central United States to Argentina—plus Trinidad, Tobago, and Grenada. It is a member of the long-nosed group of armadillos, which are virtually hairless and have broad and some flattened shells, muzzles that are quite blunt-ended, and large and upstanding ears. The nine-banded armadillo has a head and body length of 17 inches (43 cm), its tail is 15 inches (38 cm) long, and it weighs up to 15 pounds (6.8 kg). Although in the central regions of its range individuals have nine bands, in the northern and southern parts many specimens have between eight and eleven bands. It is probably the most gregarious species, as several animals may share a burrow, but it has nonspecific habitat and environmental preferences, being equally at home in forests and grasslands, in the tropics and in the temperate north. It spends the daylight hours asleep in burrows of its own

Nine-banded Armadillo *A nocturnal relative of the anteaters, the armadillo has good senses of smell and hearing, but does not see too well. It hides during the day in burrows that it digs with its powerful legs and strong claws.*
Photo: Courtesy USFWS, John and Karen Hollingsworth

digging, which is where it also raises its young. Litters always contain four babies, which are always of identical sex, as the egg divides after fertilization.

Hairy Armadillos (*Chaetophractus*)

There are two species of hairy armadillos, the small hairy armadillo (*C. vellerosus*) and the large hairy armadillo (*C. villosus*), but there is disagreement over their taxonomy and they may be races of a single species. There is certainly a considerable size difference between them, however, the smaller one weighing about 2 pounds (900 g) while the larger species is double the size. They both live on the pampas and in the mountain valleys of southern South America, from northern Paraguay to central Argentina, where they hide during the day in long burrows dug in sandy soil. They are large-eared and many-banded animals, with up to eighteen bands of which almost half are movable, and they are the hairiest of the armadillos—an adaptation for their temperate environment. Hair grows from the bands between their scutes, and they have long, pale hairs on the belly which project below the shell. In addition, they store fat and survive on metabolic water during inclement weather. They locate underground food by forcing their heads into the soil and then rotating their bodies to make a hole.

Sloths

The three species of sloths share the same order as the armadillos and the anteaters, and for many years were known as edentates after their order *Edentata*— the toothless ones. It really was an unsuitable name, as only the anteaters lack teeth; the sloths and armadillos are quite well-toothed with continually growing premolars and molars, although no incisors or canines. Their teeth are brown as they lack enamel, and their front premolars are pointed and resemble canine teeth. However, their order was recently renamed *Xenarthra*, based on their common characteristic of articulations between the lumbar vertebrae, which are known as xenarthrus vertebrae. Within their new order they now occupy the family *Megalonychidae*, another recent change from the long-used *Bradipodidae*. The sloths' most well-known characteristics are their sluggish behavior and their habit of spending their lives hanging upside down. They are nocturnal and are very slow and definite in their movements, rarely moving from the site where they have chosen to rest during the day, which is generally in a tree fork that supports their back, rather than hanging by all four limbs, and it is often in a very exposed position. The sloth's arms are longer than its legs, so its head is always lower than its backside when it hangs by its long claws, not its toes, from horizontal branches. Sloths are totally arboreal, and when they come down to the ground, which they do once weekly to defecate at the base of a tree, they crawl laboriously, although they are good swimmers. Their lazy appearance is deceptive, however, for they defend themselves by slashing with their forefeet and sharp claws, and can give a powerful bite.

The sloth's body temperature is not completely internally regulated and therefore constant, and fluctuates with the air temperature—an almost poikilothermic or "cold-blooded" physiology more akin to reptiles than mammals. As they have not evolved any hibernation or migratory strategies for survival during lowered temperatures, the sloths are restricted to the lowland tropics. The two-toed sloths have the lowest and most variable body temperature of all mammals, ranging from 75.2°F (24°C) when resting to 91.4°F (33°C) when "active," lower even than the more primitive, reptilian-like echidnas which average 86°F (30°C), and they have a low basal metabolic rate. Their eyes face forward and although their eyesight is not very good they may have a degree of color vision. Their sense of smell is well developed but they have poor hearing, and their ears are barely visible beneath their long, shaggy hair.

Two-toed Sloth (*Choloepus didactylus*)

The two-toed sloth lives in the tropical forest zone of western South America from Colombia to Brazil, east of the Andes. It has a head and body length of 30

Two-toed Sloth *With the lowest and most variable body temperature of all mammals, the sloth has an almost "cold-blooded" or reptilian physiology and a low metabolic rate. It sleeps by day in the fork of a tree, often in full view of potential predators.*
Photo: Steffen Foerster, Shutterstock.com

inches (76 cm) and weighs 17 pounds (7.7 kg). It is grayish-brown with darker shoulders and a paler face; but frequently has a greenish tinge to its coat due to the presence of algae. Its forefeet have only two digits, which are enclosed in skin right up to their long claws, but on the hind feet there are three sharp, hooked claws. The two-toed sloth's forelimbs are longer than the hindlimbs; it has a wide face with well-spaced and forward-pointing eyes, and individuals vary in their number of neck vertebrae—from six to eight. It is a more active animal than the three-toed sloth, and moves between trees more often. Its single baby is born after a gestation period of almost six months, and clings tightly to its the mother's abdomen. Although it is basically a folivore, this species has a wider-ranging diet than the three-toed sloth and also eats ripe fruit. Jaguars and harpy eagles are its major predators, and possibly the Guiana crested eagle.

Three-toed Sloth (*Bradypus tridactylus*)

This sloth is a rain forest animal with a distribution from Honduras to northern Argentina, where its major predators are the harpy eagle and the jaguar. It differs in several ways from the two-toed sloth. It has three toes on all four limbs, and its forelimbs are just slightly longer than the hindlimbs, and not as long as those of the two-toed sloth. It is a strict folivore that eats the shoots and young leaves of a variety of forest trees, and the difficulty of replacing its normal diet outside its natural range makes it an unsuitable animal for zoological gardens. It has always been linked with the cecropia tree, the leaves of which are believed to form a major part of its diet. The three-toed sloth has a coarse and shaggy gray coat, with a pale forehead and dark rings around its eyes. To shed water the hairs of its coat point downward when it is hanging from a branch, in the opposite direction to most mammals. It has a small, rounded head with small eyes and ears, and it can turn its head through an arc of 270 degrees.

Notes

1. As a source of food for the vertebrates, as opposed to the more plentiful microscopic forms such as bacteria.

2. The giant anteater (a diurnal animal) eats approximately 30,000 ants daily.

3. A charm believed to have magical powers.

8 Nocturnal Chisellers

There are 1,800 species of rodents, over half of all the mammals, and most are nocturnal. Although they vary considerably in size, shape, color, and habits, their name is unfortunately always associated with rats and mice, long, naked tails, and a certain revulsion for their way of life. While the wild "vermin" species may indeed deserve their reputation, people and rodents have a long association. Domesticated rats, mice, and gerbils are common pet animals, "cuddly" hamsters are even more popular, and the chinchilla and coypu are major suppliers of pelts to the fur trade. Many people the world over do not have the westerners' distate for rodents, even those with rats' tails, and rodents of all kinds are eagerly sought for food, not just by primitive hunter-gatherers in the rain forest, but by people in towns and villages, in markets and restaurants. In Andean towns roasted guinea pigs are regularly available, while in Asia rice rats and cloud rats are eaten, and in Africa the giant pouched rat and the grass cutter rat are regularly sold or bartered in village markets. So acceptable are rodents for the table that several species are now being considered commercially viable for farming, to supplement the dwindling wildlife. The only aspect of rodent use, terminal or invasive, about which we have few qualms is their utilization in the laboratory, for almost 95 percent of all animals used in biomedical research to assist the human race are rodents, mostly rats and mice.

Rodents generally, and therefore the nocturnal species also, vary considerably in appearance and size, ranging from the typical mouse-like animals to large, sharp-quilled porcupines and aquatic beavers. Although the members of most mammalian orders, the cats, dogs, and antelopes, for example, are similar and instantly recognizable, rodents show such great diversity of shape, size, color, and coat that early biologists had difficulty classifying them until they looked at their teeth, which leave no doubt about their relationship. They all have a pair of large incisor teeth in each jaw, usually orange or brownish in color. These "chisel-teeth" are the perfect gnawing tools, open-rooted and continually growing so they can never wear out—which are called hypsodont teeth. In conjunction with their specialized jaw muscles these teeth are tough enough to bring down a tree or chew metal. The rodents lack canines and premolars so there is a gap called the diastema between their cutting incisors and their chewing molars. In several nocturnal species, such as the pacarana and the chinchilla, the molars are also hypsodont, with open bases, and continue to grow throughout the animal's life. In other species, including the New World tree porcupines and the coypus, the molars are rooted and stop growing when the animal is mature, such teeth being termed brachydont.

Gnawing is not accomplished by the top and lower jaws acting together in a pincer-like movement, but by holding the top incisors against the object while the bottom jaw moves its incisors forward and upward against it, meeting the upper incisors, with the friction keeping them sharp. The molars do not meet while this is happening, allowing the animal to gnaw without abrading its back teeth or swallowing the gnawed material. The reverse happens when a rodent chews, for the lower jaw moves backward slightly, opposing the molars but not the incisors, so it can chew without grinding down its chisel-teeth. The name rodent is derived from the Latin *rodere*, "to gnaw," and the lack of hard foods or friction[1] can result in their incisors growing out beyond the lips and curling back like a boar's tusks. Apart from such tooth overgrowth or malocclusion, nothing interferes with their chewing. Their lips can close behind their teeth so they chew whether underwater like the beaver or underground like the naked mole rat. In hardness tests the rodent's tooth enamel proved tougher than steel.

Rodents are adaptable and prolific animals that convert vegetation into meat very quickly, and most species are highly productive, having large and frequent litters. Some can take advantage of good times to "explode" in numbers, which in turn assists other animals higher in the food chain to also have a good breeding year. Predators are continually seeking them, and rodents provide many small carnivorous animals with the bulk of their food. They must be on guard continually against attack by hawks and owls, snakes and lizards, small cats and dogs, genets, mongooses, and many others, and even omnivorous animals such as hedgehogs, pigs, and armadillos, which would eat their young. The wood mouse or long-tailed field mouse (*Apodemus sylvaticus*), a tiny species from Eurasia, is typical of the fecundity and vulnerability of the small rodents. Its frequent litters contain up to nine young after a gestation period of just twenty-five days, but they are sought by weasels, owls, cats, and shrews and the few that survive to adulthood can expect a lifespan of only twelve months. Wherever they live, the rodent environment is a hostile one, and it is not surprising that they are wary to the extreme. The juvenile mortality rate of most rodents has been calculated at about 90 percent, and to ensure that enough survive from each litter to perpetuate the species they have evolved many behavioral adaptations.

The most important rodent behavioral adaptation is the adoption of nocturnal habits, primarily to reduce the risk of predation, and even the vermin species, which now live in close association with mankind, are seldom seen during the day when people are about. The desert-dwellers have another reason for nocturnal activity—to avoid the searing heat of the midsummer day. However, there are still the hours of daylight to consider; hiding by day is the natural complement of nocturnal behavior,

Wood Mouse *Probably the Palaearctic Zone's most common mammal, the tiny wood mouse is rarely seen. At night it is active in woodland and hedgerows, and during daylight it sleeps underground in an extensive burrow system.*
Photo: Steve McWilliam, Shutterstock.com

and is a very important feature in the survival strategies of the rodents. After exposure at night to nocturnal predators they cannot risk a second round of heavy predation by the day shift. For most terrestrial rodents hiding during the day means underground, usually in a burrow of their own digging, where they sleep, give birth, and raise their young. They also store food in their burrows, estivate there to escape the hottest days of summer, and hibernate there to avoid the coldest days of winter.

Burrows provide a microclimate that differs remarkably from the desert above. The midsummer temperature of kangaroo rat tunnels in the Mohave Desert has averaged 86°F (30°C) despite the intense heat outside. This has greater importance than just being comfortable at home, as these animals have succumbed to temperatures over 95°F (35°C). At night the situation is reversed, with the burrows retaining their warmth while the desert cools considerably.

In addition to their behavioral adaptations evolution has also equipped the rodents with highly developed senses to aid their survival. Most of the knowledge of their senses comes from studies of the domesticated Norway rat and house mouse in the laboratory, plus observations from animals kept as pets. These studies prove that rodents have acute senses of taste, touch, hearing, and smell, but do not have very good eyesight. Rats and mice are in fact short-sighted, with sharp close vision but poor ability to focus on objects beyond 48 inches (1.2 m), although they are sensitive to movement up to 50 feet (15.2 m) away. They are believed to have a degree of color vision, which of course is useless at night anyway, but in low light levels their other acute senses more than compensate for their poor vision. The desert rodents have enlarged, almost spherical lenses and concentric retinas, which gives them horizon-scanning sight but with poor powers of accommodation. They rely on this sense more for warnings of approaching danger, and when searching for food depend upon their sense of smell. Their large black eyes glow ruby red at night as light is reflected back by the tapetum.

The rodent sense of taste is very highly developed, and rats and mice can apparently detect contaminants in their food at levels as low as 0.5 parts per million, which is why they often reject poison bait. They also have a highly developed sense of smell, receptor cells in the nasal epithelium having cillia which project into the mucus of the nose on one side and into an olfactory bulb on the other. Rats have 50 million receptor cells in each nostril compared to the 6 million in the human nostril. Smell is used to locate food, traverse their pathways, and for communication, for they mark objects with urine and glandular secretions that contain pheromones for communication between conspecifics. From these pheromones they can recognize other rats and familiar odors along the regular pathways, and can learn several things quickly about strange rats from their scent, such as their sex, their readiness to mate, and their "character"—whether they are weak or strong individuals. The rodent sense of touch is also acute, their sensitive tactile vibrissae moving constantly as they check out their environment. They are contact animals, preferring to move along a pathway that has a stationary object, such as a wall, along one side.

Rodents also have excellent hearing, and many have large outer ears, those of the long-eared jerboa (*Euchoreutes naso*) being relatively longer than a hare's ears. Species with small ears have greatly enlarged tympanic or auditory bullae (the prominence below the opening of the ear) which contain air of high humidity that possibly prevents dehydration of the fluid-filled middle ear, but certainly improves hearing sensitivity. These bullae are so large in some species, such as the three-toed dwarf jerboa (*Salpingotus kozlovi*), that their heads seem far too large for their small bodies. In Ord's kangaroo rat (*Dipodomys ordi*) the middle ear can apparently detect the acoustic frequencies of an owl's wing-beats and a striking snake, which sometimes helps them evade these predators.

Some of the Species

Rat-like Rodents (Myomorpha)

Norway Rat and Black Rat (*Rattus*)

These are very similar cosmopolitan animals that are frequently mistaken for each other. The black or roof rat (*Rattus rattus*) is a sleek, black rodent with a gray belly and a naked tail that is as long as its head and body, and measures about 9 inches (23 cm). The black rat has a great sense of balance, and is an agile climber that lives in trees and also in the attics and lofts of buildings to which it can gain access along power lines. It eats fruit and nuts, birds' eggs and nestlings, and raids stored foods in granaries and warehouses, travelling over a mile (1.6 km) if necessary from its untidy nest to a favored feeding place. The black rat is most common in the tropics while the brown rat is more plentiful in temperate regions. It probably originated in the Far East and spread naturally across Asia, and was then brought to Europe during the Crusades and crossed the Atlantic Ocean with the Pilgrim Fathers early in the seventeenth century.

The Norway or brown rat (*Rattus norwegicus*) is more common than the black rat and is dominant over it where their habitats overlap. Its dramatic spread around the world has in places been at the expense of the black rat, which in some regions has been declared an endangered species, although it is unlikely to ever be the subject of conservation programs. It has a grayish-brown coat and is slightly larger than the black rat, measuring 10 inches (25 cm) in length, but with a shorter tail, which only reaches the rat's ears when pulled back over its body. It is primarily an animal of farmyards and urban areas, especially garbage dumps, granaries, dock areas, and slaughterhouses, wherever there are stored food products or waste. It is a more terrestrial animal than the black rat and usually lives in long, branching tunnels. The Norway rat's origins are also believed to be in the Far East, perhaps China; it arrived in Europe in the sixteenth century and reached North America at the end of the eighteenth century.

Both species are highly reproductive, probably the most prolific of all the rats, producing litters of up to ten young several times annually. They are parasites that have caused great loss of human life, loss of wild animals in lands to which they have been introduced, and much damage to stored foods. They have transmitted many diseases including bubonic plague, leptospirosis, and tularemia, and have been directly responsible for more loss of human life in the past millennia than all the wars combined. They are complete omnivores, which means they eat literally anything, animal or vegetable, and sometimes mineral too; plus many things which we do not eat such as hide and soap. They have a special fondness for animal life, in any form they can overcome, and hunt at night. They attacked my roosting chickens in Wales; lamed a zoo elephant by chewing its feet each night until its keepers realized the cause, and have gnawed on wounded soldiers lying immobile in their hospital beds in India.

House Mouse (*Mus musculus*)

Like the vermin rats the house mouse is now found throughout the world in association with people, except in Antarctica. It probably evolved in Eurasia, where it still lives in the wild, in extensive burrow networks that have several exits and which include chambers for storing food, for sleeping, and for raising the family. They likely reached Europe and then Britain from western Asia during the glory days of the Roman Empire. Where it is not disturbed, the house mouse may appear during the day, but in its close association with people it is a purely nocturnal animal that occupies farm buildings, human dwellings, warehouses, and granaries, and wherever there is a source of food. It is rarely seen during the day unless disturbed when a sack of grain or a sheet of plywood is moved. It is especially destructive to foodstuffs as it contaminates far more than it eats.

The house mouse is a grayish-brown animal with paler underparts, only 3 inches (7.5 cm) long with a scaly tail of about the same length. Highly reproductive, it can breed when only fifty days old, has a gestation period of just twenty days, and the babies (up to twelve in a litter) are weaned when only twenty-eight days old. Like the black and brown rats, the house mouse eats virtually anything, but does not have the rat's propensity for killing animals, other than insects. Zoonoses—diseases that mice can transmit to humans—include leptospirosis and salmonellosis from contaminated products, and hantavirus from aerosols of urine and fecal material.

Giant Pouched Rat (*Cricetomys gambianus*)

An enormous rat, with a head and body length of up to 18 inches (46 cm) and weighing over 3 pounds (1.3 kg), the pouched rat lives in thickets and forests throughout Africa south of the Sahara. Its name refers to the large cheek pouches for packing food and not marsupial-type stomach pouches for carrying young. Pouched rats have short and thin grayish-brown fur with pale underparts and white patches on the sides of the face, and a long, rat-like tail, but despite their appearance they soon become very tame and make excellent pets. They are mainly terrestrial, but climb and swim quite well, and dig their own burrows or live in natural crevices and the abandoned holes of other animals. Pouched rats are mainly herbivorous but also eat snails and land crabs in addition to plant matter. They are not sexually mature until the age of five months, which is old for a small rodent; their gestation period is thirty days and their litters average four young. They breed continuously throughout the year and in captivity have proved to be quite social, with several animals living compatibly together. These qualities have encouraged investigation of their farming as a cottage industry in West Africa, where they are a favored source of bushmeat, but the wild stocks are dwindling and can no longer support regular harvesting. Farmed rats would be more acceptable to the villagers than domesticated rabbits, which are the obvious herbivorous alternative. Pouched rats have tiny eyes and poor vision, relying more upon their keen senses of smell and hearing.

Edible Dormouse (*Glis glis*)

This unusual rodent occurs naturally in continental Europe, and has been introduced and is now established in southern England. It is a nocturnal woodland animal that eats nuts, fruit, acorns, birds' eggs and nestlings, and causes great damage in orchards and vineyards. The edible dormouse has short, thick silvery-gray to brownish-gray fur, with a white belly and a dense, bushy tail. Adults naturally weigh about 6 ounces (170 g), but the ancient Romans fattened them with chestnuts and acorns. They were kept either in enclosures called gliraria or in small earthenware pots, and when ready for the table they were triple their normal

Giant Pouched Rat *Pouched rats are active at night in the forested regions of sub-Saharan Africa, filling their cheek pouches with food to carry back to their nests where they spend the daylight hours. Their senses of smell and hearing are well developed, but they have poor vision. Overhunting for food has seriously reduced their numbers throughout their range.*
Photo: Clive Roots

weight and so fat they could barely move. They were apparently roasted very crisp, and were a food of the nobility and the rich, such a luxury that at one time their consumption by the general populace was banned. The ability to store so much fat stems from their natural hibernation physiology, providing food for their dormancy from October to April when they sleep in tree holes or underground, with several animals often curled up together. Edible dormice are quarrelsome little animals, which in the wild mate and then go their own way; captive pairs have lived together, although females did not allow males near the nest box while the babies were small. Their nocturnal senses are highly developed, with good vision, hearing, and smell, and very tactile vibrissae.

Desert Jerboa (*Jaculus jaculus*)

Jerboas are natives of the desert zone from North Africa through Asia Minor and the Arabian Peninsula to southern Iran. They are similar to the North American kangaroo rats, but are even more kangaroo-like as their tails form a tripod with their legs and support the body in a similar manner. The desert jerboa is one of the smaller species, weighing about 2 ounces (57 g), with a head and body length of 6 inches (15 cm) and a tufted tail half as long again. It is a nocturnal burrower that digs its tunnels into hard ground, and these have an escape hatch with just a thin covering of crusted sand through which they burst out when they are alarmed. Desert-dwellers eat them, following their tracks in the sand and then probing down with sticks to waken the sleeping animal and flush it out of its bolt hole. Jerboas are generally solitary, but have been kept in small groups that have even slept huddled together. In the wild desert jerboas are believed to hibernate during the colder months, as specimens in a torpid-like condition have been found then, and they also sleep for days during long hot periods, in the form of summer dormancy known as estivation.

Common or Black-bellied Hamster (*Cricetus cricetus*)

The largest hamster, a stocky, guinea pig–sized animal that reaches a length of 12 inches (30 cm) with a tail just 2 inches (5 cm) long and weighing up to 2 pounds (900 g), the common hamster is light brown above and has a black belly and white patches on its cheeks and sides. Albinos and melanistic individuals are often seen. It ranges from central Europe to central Asia, although it is now very rare in the western parts of its region, and is mostly crepuscular in its habits. The common hamster lives on the open steppes and along river banks, in burrows of varying depths, but reaching at least 6 feet (1.8 m) deep to escape the frost during its winter hibernation. Like the other hamsters it stores food for the winter, and wakes every few days to feed. It needs a large supply to maintain it for several months; storage chambers holding 200 pounds (90 kg) of seeds have been excavated, and in fact purposely raided by people to steal its stores. As a result of hibernating all winter, the common hamster is a seasonal breeder, giving birth during late spring and summer, and possibly having no more than two litters annually. It is a solitary and aggressive animal, with the males entering a female's territory only for mating and being driven out soon afterward.

Squirrel-like Rodents (Sciuromorpha)

Giant Kangaroo Rat (*Dipodomys ingens*)

With its well-developed hindlimbs and bounding gait the giant kangaroo rat, and others in the genus *Dipodomys*, resemble miniature kangaroos. It is the largest species, weighing 6 ounces (170 g), whereas the other species average only 2 ounces (57 g), and has a restricted distribution in central California. Kangaroo rats have hairy tails, these being distinctly tufted at their ends in several species, and fur-lined cheek pouches in which they gather seeds in large amounts to store in their underground chambers. A cache of seeds weighing 12 pounds (5.4 kg) was discovered in one burrow. They have often been kept as pets and also have figured prominently as laboratory animals in view of their ability to survive without water. In the wild they obtain some of their fluids from succulent plants, but have also evolved renal mechanisms that actually conserve water by concentrating their urine to a density far greater than that of sea water, and seventeen times the osmotic pressure of their own blood. In comparison man can only concentrate his urine to four times the density of his blood, and like all other animals would increase the salt content of his blood, hence his thirst, if he drank sea water. They also considerably reduce the moisture content of their feces, and they economize further on moisture loss by neither sweating nor panting. Kangaroo rats can also produce their own water through body metabolism, and can live on a diet of seeds alone. They select seeds for storage based on their water content, and store dry seeds in humid burrows to increase their moisture. During very hot weather they attempt to increase their consumption of insects, which have a high preformed water content.

Springhaas or Spring Hare (*Pedetes capensis*)

Certainly one of the most attractive of all rodents, the springhaas is another kangaroo-like animal the size of a small hare, with short forelimbs and powerful hind legs, but with a long and bushy tail. It also stands like a kangaroo, resting tripod-like on its hindlimbs and tail. Its long, soft fur is tawny reddish-brown on the back with underparts of buffy-white, and a white line in front of the thighs extends up toward the spine. The tail is dark brown or black with a thick brush at the end. The springhaas lives in the southern half of Africa, from Angola and Kenya to Cape Province, in arid and semiarid regions. Adult males have a head and body length of 17 inches (43 cm) and their tail is slightly longer; they weigh almost 9 pounds (4 kg). They are very social animals that live in burrow systems, with an adult pair and their young occupying separate tunnels within the complex. They use their burrows for shelter, raising their young, and to escape predators, and leap high when they venture outside to foil lurking predators. They are nocturnal, but have reversed their activities when kept as house pets and becoming tame. Springhaas eat bulbs, fleshy roots, seeds, plants, locusts, and beetles, and store food for the cold winter months, but they do not hibernate. They are favorite "bushmeat" animals throughout their range, and are hunted at night when their eyes glow red in torchlight.

North American Beaver (*Castor Canadensis*)

This familiar large rodent, weighing up to 80 pounds (36 kg), has a short and dense coat of soft and glossy dark-brown underfur protected by guard hairs. It is mainly nocturnal and usually fells trees and drags leafy branches to its underwater store under cover of darkness. Well adapted for a semiaquatic existence, the beaver has small eyes protected by a nictitating membrane, valvular nostrils, and short ears that can be closed underwater. It has short legs and large hind feet with webbed toes; the flat and naked scaly tail is 16 inches (40 cm) long and 5 inches (12 cm) wide, and the sexes are alike in size and color.

The beaver is a native of North America and northern Mexico (there is another species in Europe and western Asia), where it dams flowing fresh water to create a deep lake where it can build its semi-submerged lodge for its winter security. It does not hibernate, and stores its winter food supplies on the lake bottom, to which it has direct access underwater from its lodge, where the young are born in late winter. The beaver has a single opening or cloaca into which the anal and urogenital orifices open, and its paired castor or scent glands discharge into the urethra and then into the cloaca, this excretion being used to mark the lodge and territory. With protection and reintroduction during the last century it is now a common animal, after having been almost exterminated during the nineteenth century to supply the fur trade.

Flying Squirrels (*Glaucomys*)

There are two flying squirrels in North America—the southern flying squirrel (*G. volans*) and the northern flying squirrel (*G. sabrinus*). The southern species, a native of the eastern United States, is distinguished by its white belly and its smaller size, weighing a maximum of only 3 ounces (85 g). The northern species lives in the northern United States and across Canada south of the tundra zone, but excluding the prairies. It is double the weight of *G. volans* and the hairs on its belly are only white at their tips. Both species are seldom seen as they are totally nocturnal and hide in natural tree crevices and hollows during daylight. The flying squirrels have a membrane called the patagium, which extends between the limbs on both sides of the body, opening out like a blanket to increase their surface area, allowing them to glide down from a high point. Direction control is provided by membrane-adjusting muscles and by angling the long, flattened tail like a rudder, but the membranes cannot be flapped in true vertebrate flight in the manner of the birds and bats. They are gregarious animals that live in extended family groups, and several huddle together for warmth in winter, remaining in their nests in semi-torpor during very cold weather. *Glaucomys* are vegetarians, their favorite foods being nuts, berries, shoots, and the sap of the sugar maple tree.

Western Pocket Gopher (*Thomomys mazma*)

Pocket gophers are mainly fossorial animals; they live most of their lives underground, where they are active day and night, but *T. mazma* differs in spending more time above ground during darkness, collecting supplies for its stores. They search for seeds, grasses, garlic, lupines, and dandelions within a small radius of their entrance hole and snip them into small pieces which they tuck into their fur-lined external cheek pouches to carry back to their stores. The pocket gopher also feeds underground on roots, bulbs, and rhizomes, and may pull plants straight down into its burrow after snipping off the roots. It is a small animal, about 8 inches (20 cm) long with a highly sensitive tail 2½ inches (6 cm) in length, and has a robust body with short smooth hair, short legs, and long foreclaws. Its color varies according to the type of soil in its habitat, and may be reddish-brown, gray, or even almost black; it has a large head and conspicuous incisor teeth that protrude beyond its closed moth. It leads a solitary life, although individuals' burrows may be so close to each other, due to their small range, that they give the impression of being colonial. When above ground pocket gophers are at risk from owls, snakes, bobcats, and foxes.

Cavy-like Rodents (Caviomorpha)

Chinchilla (*Chinchilla laniger*)

Chinchillas were almost exterminated in their natural habitat—the higher elevations of the central Andes from southern Peru to northwestern Argentina—due to the great demand for their luxurious pelts early in the last century, and are now restricted to the mountains of northern Chile. Only their eventual breeding on fur farms to meet the demand relieved the pressure on the wild populations. They are still very important animals in the fur industry, and their domestication since they were first brought to the United States in 1923 has resulted in the production of many color forms in addition to their natural silver-gray. They are very hardy animals, their dense and soft fur protecting them from the low temperatures of high altitudes, but they are very sensitive to heat and high humidity, and temperatures above 25°C (77°F) may cause heat stroke, with elevated body temperature, panting, immobility, and their eventual death. They have a head and body length of 14 inches (36 cm), their tail is about 5 inches (12.5 cm) long, and they have large ears and large black eyes with vertical pupils in contraction. Chinchillas are social animals that live in small colonies, and they breed twice annually, averaging six young per litter; their babies are born fully furred and with their eyes open.

Pacarana (*Dinomys branickii*)

A large paca-like rodent, the only member of its genus, the pacarana has a coarse dark-brown coat with rows of close white spots almost forming continuous lines along its back from the shoulders to the rump, with shorter and more open rows of spots on its sides. It has a large and broad head, small ears and eyes, and short legs with powerful claws. Unlike the similar spotted paca, it has a tail that is about 8 inches (20 cm) long, and is slightly larger, with adult males weighing up to 33 pounds (15 kg). The pacarana is a very rare animal, seldom seen in zoos and always in great demand. It is now also scarce in its native habitat in the valleys and lower slopes, up 6,500 feet (2,000 m), in the tropical Andes, from Venezuela to Bolivia. Captive animals have been both diurnal and nocturnal, but in the wild they are nocturnal and enlarge natural holes in which they hide during the day. At night they rely on their senses of smell, taste, and touch, because their tiny eyes and small ears do not provide high-quality vision or hearing. Their major predators are wild cats—jaguars, ocelots, and cougars. Pacaranas sit upright to eat, holding food in the forepaws, and can walk bipedally for short distances.

Pacaranas *A cavy, and therefore a relative of the guinea pig, the large—and now very rare—pacarana lives on the slopes of the northern Andes, where it sleeps during the day in natural crevices, and appears at dusk to graze and browse on grasses, leaves, and bark, and search for fallen fruit and berries.*
Photo: Clive Roots

Coypu (*Myocastor coypus*)

A large, rat-like South American rodent, in which adult males may weigh 17 pounds (7.7 kg), the coypu is now established in several countries, including England, the southern United States, and Russia, after escaping from fur farms or being deliberately released. Escapees in England years ago entered the sewer systems and were called giant sewer rats. The coypu is a major commercial species known in the fur trade as nutria, the Spanish for otter, and has also been widely used as a laboratory animal. Consequently, it is now considered domesticated and is available in several color mutations in addition to its natural coat of grayish-brown with golden guard hairs. Coypu are mainly nocturnal, semiaquatic animals with webbed feet, and they swim and dive very well. They are primarily herbivorous in the wild, eating water plants and raiding riverside crops or grazing along the river banks, where their burrowing has caused considerable damage. They also eat molluscs and scavenge on dead fish. The coypu has well-developed night vision, its eyes having vertical pupils in contraction during daylight, and its senses of smell and hearing are also quite acute. The female's mammae are situated high on her sides so she can feed her youngsters while swimming.

African Crested Porcupine (*Hystrix cristata*)

The largest and most impressive of the porcupines, this animal's body is covered with long, hairy spines and sharp quills, and its head is blunt with the eyes set back on its sides. It occurs throughout Africa excluding the equatorial forest zone, and has a very limited distribution in southern Europe. It has a relative in central and southern Asia—the Indian crested porcupine (*H. indica*). An adult of the African species can weigh up to 66 pounds (30 kg) and reach a length of 2 feet 8 inches (81 cm) with a 5-inch-(12.5 cm)-long tail that has open-ended quills. The body quills are used in defense when the porcupine raises them and rushes backward at an aggressor, and as a warning when it rattles them, when some may fall out, resulting in the fallacy that it can shoot its quills. Crested porcupines are nocturnal and hide during the day in caves, in wide rock crevices, or in burrows that they dig themselves or borrow from aardvarks and even hyenas. Smell is probably their most acute sense, as their eyes and ears are small and vision and hearing are therefore unlikely to be well developed. They are social animals that breed throughout the year, but females prefer to give birth in a separate den. These large porcupines are mainly vegetarians but they also eat carrion, and chew on bones for the calcium. They are wasteful feeders that cause great damage to crops.

North American Porcupine (*Erethizon dorsatum*)

The North American porcupine is a heavy-bodied, short-legged animal with a short, thick tail and is covered with medium-length quills and long, blackish hair-like spines on the back and nape. When adult it may weigh 33 pounds (15 kg). It is a native of forested areas in Canada and the western United States, and is a very hardy animal that dens in hollow trees and caves, but does not hibernate; it is out on the coldest nights of winter, eating mainly the inner bark of trees. It has a very long digestive tract, which may measure 33 inches (850 cm), almost half of which is small intestine where bacteria decompose its high cellulose intake. During the summer months it extends its diet with leaves, bark, and berries. It sleeps by day in a tree hollow or among boulders, and has few predators, except for the fisher (*Martes pennanti*), a large relative of the mink, which has learned how to safely flip the porcupine over to attack its unprotected belly. *Erethizon* has poor vision, but good senses of smell and hearing. Females are sexually mature when two years old but males are believed unable to breed until they are three; they are seasonal breeders, mating in the fall and producing a single youngster between April and June, after a gestation period of 215 days. The newborn, well-furred baby is very precocial, has its eyes open, and is able to climb and eat solid foods shortly after birth.

Note

1. They may also grind their teeth when they are not eating, to prevent overgrowth and to keep them sharp.

9 The Night Hunters

The night hunters are the carnivores[1] or predatory land mammals, members of the order *Carnivora*, although the name is misleading as several members of the order have diverged and rarely eat meat, so they certainly cannot be considered hunters. Taxonomically, membership in the order is based upon common evolutionary relationships[2] and certain skull characteristics, not an animal's current diet, and their evolved feeding behavior has resulted in three types of carnivores—the carnivorous, omnivorous, and vegetarian carnivores. The cats, dogs,[3] and hyenas are carnivorous and are hunters, as are the weasels and polecats, genets and linsangs. The omnivorous forms such as two of the bears, the skunks, badgers, and civets are not hunters, and neither are the fruit-eating kinkajous and olingos, but they are all nocturnal and are therefore included here, as aberrant carnivores.

The hunters have perfected mechanisms for seeking, seizing, and killing their prey, and for digesting and utilizing its energy. While they are all experts in their own right, equally successful at catching and killing other animals, their hunting methods differ considerably. The large cats are mainly stalkers and ambushers, using little energy until their final burst of power, whereas the small cats are pouncers, locating their small rodent prey by noises or movement in the grass. Several dog species are pursuers, social pack animals that exhaust their prey after a long chase; the joint efforts of several snapping animals are needed to bring down large victims. The weasels and polecats are also persistent pursuers, although solitarily; they follow a scent trail and their terrified prey even down rabbit holes, whereas the arboreal genets pursue their prey through the treetops.

■ WILD CATS

Cats are members of the order *Felidae* and zoologists consequently call them felids. Only the smaller species actually go by the name cat, however, such as leopard cat, jungle cat, and fishing cat, the larger species having more distinctive names like lynx, leopard, cougar, lion, and tiger; but "cats" is a perfectly acceptable collective name for them all. They certainly look alike and are immediately recognizable, differing mainly in size, from the small sand cat that weighs 4 pounds 4 ounces (2 kg) to the Siberian tiger at 683 pounds (310 kg); and in color, with either plain, spotted, or striped coats. They also differ in their ability to roar. In the larger cats of the genus *Panthera*, excluding the snow leopard, the hyoid[4] has failed to develop completely and has elastic cartilage between bones, which allows them to roar and to purr only when exhaling. In the smaller cats of the genus *Felis*, the hyoid is completely ossified—the natural process of bone formation—so they cannot roar, but purr when both inhaling and exhaling. Excluding the lion, cats are solitary and very territorial animals, the nocturnal species being skulkers and ambushers, adapted for a short, rushing attack, and stealth is their most important asset. Broad footpads act as muffles to soften their tread and they appear to glide along as they move the limbs on either side of the body together.

Cats are considered the ultimate hunters, but their methods of execution vary. Lions are the only gregarious hunters, which cooperate to drive the prey toward other pride members lying in ambush. Solitary hunters like the cougar and the leopard leap onto their victim from an overhead branch or rock ledge, or seize it after a brief chase. The tiger lies in ambush in thick vegetation or stealthily approaches its unsuspecting victim until it is close enough for the final attack. The outburst of energy required to power this last killing rush is unequalled in the animal kingdom. The tremendous power is mobilized by the adrenalin circulating in the blood stream, which produces a sudden flash of oxidation in the nervous system and the brain. Stimulation of the adrenal glands, nervous system, and brain increases respiration, heartbeat, and the blood's circulatory rate, sending sugars into the bloodstream from the liver. Such an explosive outburst is very exhausting, but when a kill is made the energy is replaced by the gorging of large quantities of meat and blood, providing the cat with protein and energy.

Cats are strict carnivores that need a pure meat and fat diet that provides twenty-three amino acids on a daily basis, unlike dogs, which tolerate carbohydrates, and diets formulated for domesticated dogs are therefore inadequate for cats. The dietary protein requirement for growth in cats is 35 percent, and as adults they have the highest requirement for daily maintenance (up to 30 percent) of all land mammals. As they have evolved a lifestyle that excludes the intentional ingestion of carbohydrates, although they may incidentally eat their prey's gut contents, cats now lack the enzyme glucokinase which is required for the digestion of carbohydrates.

Although largely nocturnal, cats may hunt during the day, which is usually an indication that the previous night's efforts were unsuccessful. They kill small prey by driving their long canines into the nape of the neck, where they penetrate the hind brain or sever the spinal cord. Lions and tigers leap onto the back or flanks of large prey such as zebra or buffalo and drag it to the ground, where it is held by the muzzle or throat and killed by strangulation. They have developed knife-like carnassial teeth, the enlarged fourth upper premolar and the opposing first lower molar, which have a shearing action for cutting meat or hide; and their tongues have sharp papillae for rasping meat. The manner in which the cat's lower jaw is hinged permits up-and-down movement only, with no ability for lateral grinding, so they cannot chew.

Vision is the cats' most highly developed sense. Their eyes are the largest of the land carnivores, and typical of hunting animals they are situated frontally, giving them binocular or 3D vision—the blending of the separate images seen by each eye to produce a stereoscopic image. This gives better depth perception and ability to judge distance, and helps them decide when to start the chase and when to pounce, greatly improving their chances of success. Their large eyes also bulge outward slightly, giving them a degree of peripheral vision, and they have the habit of staring unblinkingly at their prey to intimdate it, and also at other cats to establish dominance.

Cats' pupils contract considerably in bright light to protect the lens from damage, to a pinpoint pupil in the large and medium-sized cats plus the lynx, whereas the reduced pupil of the wild cat, like the domestic cat, is a vertical slit. Light penetration can be further reduced by partly closing the eyelids and thus reducing the length of the slit. Extreme dilation of the pupils at night allows full use of the available light, and the presence of a tapetum increases their powers of sight as it reflects light back through the eye. Acting like a mirror, the tapetum replaces the light-absorbing layer behind the retina of the diurnal animals and produces the "nightshine" of cats' eyes caught in a beam of light.

In most carnivores, and particularly the cats, the eyes are situated in open orbits as the bony socket does not form a complete ring around the eye. This allows the jaw to open wider, with greater capacity for seizing, holding, and crushing bone. Although closed orbits provide the best protection for the eye, in the carnivores a degree of protection is offered by the bony arch that extends from the orbit to the back of the skull. Tigers have the most open orbits and therefore the widest gape. As a result of domestication the orbit of the house cat is beginning to constrict in the center and now resembles the dog's orbit, where a bony ligament covers the rear of the eye socket. When food is short or when they are very sick, cats have a sunken hollow-eyed appearance due to the loss of the fat which is normally stored in the socket beneath the eye.

The cat's large, upright ears are extremely accurate sound-locating organs, well endowed with muscles that can rotate them 180 degrees independently of each other to assist in the detection and especially the location of sounds. Their hearing is far more sensitive to sounds of higher pitch than the human ear, and their range is 30 Hz to 65 kHz (compared to the dog's 20 Hz–40 kHz and the human 20 Hz–20 kHz), a hearing range attuned to the sounds of the small rodents that are the favored prey of the smaller cats. When hunting in the dark, and especially in thick forest, it is likely that hearing and smell are initially a cat's most important senses, followed by sight as it gets within striking range. In grassland species such as the lion and perhaps the caracal, sight obviously has a more important role at the onset of the hunt at dusk.

Cats are highly sensitive animals and touch plays an important role in their daily life. Whiskers or tactile vibrissae are very sensitive to air movement and move in response to the smallest changes in their environment, sending impulses to the

African Lions *At dusk on the African savannah, four juvenile lions accompany a lioness at the start of the night's hunt. Herbivores settling down for the night—such as impala, zebra, eland, and gnu—are all potential victims.*
Photo: Photos.com

brain for analysis. They have sensory organs called proprioceptors at their ends, which aid in detecting nearby prey. The tiniest change in air movement enables a cat to locate its victim, and it can extend its whiskers forward to their fullest when an animal just touching a whisker is quickly seized. Whiskers help the cat navigate in darkness, and if they touch the sides of a hole the cat knows that its body will not go through. Cats shed their whiskers, one or two at a time, and it is very upsetting for a cat to have its whiskers cut. In addition to its whiskers a cat's face and front paws are its most sensitive places, with high concentrations of nerve cells. The paws are sensitive to pressure, and can detect the slightest vibrations. A cat's coat and skin are also highly sensitive to temperature changes, receptors there sending nerve impulses to the brain. When domesticated cats are petted their nervous system responds by lowering the heart rate, relaxing the body, and stimulating the flow of stomach acids, thus improving digestion. Kittens, which are blind for ten days after birth, move around by touch, and the heat receptors in their noses, which can detect temperature variations of 1°F (0.45°C), help them locate their mothers.

Cats have a good sense of smell, and it is an important health factor for them, because when their smell is compromised—due to the stuffiness of a cold—they usually do not eat as they smell food before accepting it. Although they do not use smell for tracking, like the dogs, a surprisingly large part of their brain is devoted to this sense. In addition, they have a vomeronasal organ in the roof of the mouth for detecting pheromones, which involves drawing in a sharp breath through the mouth while making the facial grimace known as "flehmen." The sense of smell is important also because cats scent-mark objects widely throughout their territory with odors produced by their anal glands and three sets of scent glands—on the face, between the toes, and under the tail. Cats' tongues are covered with papillae, or small barbed hooks, and they have a very rough lick, but taste is not considered a highly developed sense, as they have only 475 taste buds (on the tip, sides, and back of the tongue) compared to 1,700 in the dog.

Although the smaller cats rest during the day in situations where they are themselves hidden and somewhat protected from predation, the larger species usually rest in full view, like the lions under an acacia or the leopard draped over a thick tree limb high above the ground.

Some of the Species

European Wild Cat (*Felis silvestris*)

The European wild cat is an animal of the forests and open heaths, generally in hilly regions throughout Eurasia and Africa, and still survives in remote areas of Britain. The north African race (*F. s. lybica*) is the ancestor of the domesticated cat. There is a great variation in size throughout its range, with males in eastern Europe having a head and body length of 29 inches (75 cm) and weighing 15 pounds (6.8 kg), but generally they are little larger than the average house cat. There is also considerable variation in their color, ranging from yellowish-gray to brownish-gray, always with paler underparts, and with striped forehead and legs and a ringed tail that has a black tip. It is one of the most solitary cats, yet captive pairs have lived amicably together and kittens have been raised in the presence of the male; however, it is a particularly secretive and nervous species, which needs complete privacy to breed successfully. Its diet includes rabbits, hares, rodents, grouse, pheasants, and partridges.

Northern Lynx (*Felis lynx*)

Lynx are animals of the forests and muskeg and are distributed across the Northern Hemisphere in the Palaearctic and Nearctic Regions. Excluding the rare and protected Spanish lynx, there is just one species across this vast region—the northern lynx—with numerous recognized races, including the Canadian lynx, Siberian lynx, and Irkutsk lynx, which are all still important commercial fur animals throughout their range. In North America the Canadian lynx (*F. l. canadensis*) ranges across Canada south of the Arctic Circle, and extends into the United States in the forests of the Pacific Northwest, the Rockies, and around the Great Lakes. Although usually considered small cats, a mature male Canadian lynx has a head and body length of 36 inches (90 cm) and can weigh up to 30 pounds (13.5 kg);

and specimens of the European lynx (*F. l. lynx*) have reached 55 pounds (25 kg) in weight. Lynx are distinguished by their thick, pale coats, tufted ears, a short tail with a completely black tip, and huge feet, which enable them to walk on soft snow. The rare Irkutsk lynx (*F.l. kozlowi*) is a very slender animal, with a pale-brown coat in which faint spots can be discerned, and its tail is tipped with black for half its length. The lynx is a very capable hunter and in North America its favored foods are white-tailed deer and mule deer fawns, cottontail rabbits, snowshoe hares, grouse, and wild turkeys. In Eurasia, it preys on red deer and reindeer fawns, chamois, musk deer, roe deer, rabbits, hares, and pheasants.

Cougar (*Felis concolor*)

The largest member of its genus, the cougar is an animal of many names, including puma, mountain lion, catamount, and painter. It is widespread throughout the New World, in a great range of habitat, from the cold and wet northwestern Canadian forests to Brazil's equatorial rain forest, plus grassland, rocky deserts, and swamps. In elevation it ranges from sea level to high in the Andes and the Rocky Mountains. The cougar has an average head and body length of 5 feet (1.5 m) and weighs 187 pounds (85 kg); but there is considerable size variation throughout its vast range, the tropical cougars generally being smaller than the northern animals. It is also variable in color, with two basic phases, a tawny-buff phase and a grayish one. The cougar is both nocturnal and diurnal and preys largely upon deer—white-tailed, mule deer, elk, and caribou in North America—plus cottontail rabbits and hares. In South America it kills white-tailed deer, brockets, agouties, capybara, pacas, occasionally young tapirs; while at the southern end of its range in Patagonia the guanaco is the major prey. Domesticated livestock are always at risk in cougar country, but it only occasionally attacks humans, most frequently in British Columbia and especially on Vancouver Island. It is a very agile cat, able to jump upward for 11 feet (3.5 m), and has good sight and hearing. In its habits it is very solitary and meets conspecifics just for mating.

Leopard (*Panthera pardus*)

The leopard has a wide range that includes sub-Saharan Africa with fragmented populations in north Africa and the Middle East, central and southern Asia, China, and within Indonesia just the island of Java. Across this vast range numerous subspecies are recognized, but hunting, poisoning to protect livestock, and loss of habitat have reduced or eliminated it in many parts of its original range, while in others it is still a fairly common animal. Leopards are very agile and powerful cats, able to drag heavy antelope high into trees to cache for later, and they are great leapers—both horizontally and vertically. Males weigh up to 200 pounds (90 kg) and females 130 pounds (59 kg). Their upper-body coloration varies from pale buffy-yellow to bright chestnut, covered with rosettes formed by black spots with dark centers, and with smaller black spots on the lower legs, head, and chest, and large black ones on their white bellies. Melanistic individuals, or black panthers, are common in the humid forests of Southeast Asia. Leopards are animals of varied habitat, including grasslands, dry scrub, open tropical woodland, northern mixed forest, and equatorial rain forest, from sea level to over 11,000 feet (5,000 m). Mainly nocturnal, they spend the daylight hours asleep in undergrowth, among rocks, or in a large tree (see color insert), appearing at dusk to hunt. In Africa their prey includes various antelope species such as impala, klipspringer, gazelle, duiker, plus giraffe calves, gnu calves, zebra foals, warthog, bushpig, baboons, vervet monkeys, jackals, hares, guineafowl, spurfowl, and ostrich chicks, plus domesticated livestock. In Eurasia they kill roe deer, axis deer (see color insert), sika deer, young sambhar and red deer, muntjac, wild boar, pheasants, and domesticated goats and sheep.

Lion (*Panthera leo*)

Lions are animals of the more open dry country including semidesert regions, savannah, thorn scrub, and open woodland. Long ago the species had a very large range, but in early historical times it was restricted to Africa, Asia Minor, and India. In the last century hunting, loss of habitat, reduced prey animals, and persecution by livestock owners reduced its range even further, and it is now restricted to Africa south of the Sahara and the Gir Forest of western India. Lions are the second largest of the cats after the tiger, with adult males reaching a weight of 550 pounds (250 kg). Their body color varies from pale buffy-gray to yellowish-red or tawny, they have a black tuft at the end of the tail, and males have a neck and throat mane that also varies in length and color, being either reddish-brown or black. They are the only social cat, living in "prides" of several males and females and their young. Prides vary in size, but average twelve animals, and the males normally allow the females to make the kill. Despite their social behavior and cooperation when hunting, lions are quarrelsome when eating, and are very aggressive to juveniles in the pride. Most of their hunting is done during darkness, and their main prey includes the normally diurnal animals (see color insert) such as impala, zebra, eland, oryx, gnu, buffalo, wart hog, and ostrich, which are caught either after a stalk and short chase, or through a cooperative effort in which the victim is driven toward others of the pride lying in ambush. They also kill young rhinos, hippos, and giraffes, but generally stay well clear of adults of those species. They are also scavengers, even of beached marine mammals along remote shorelines such as Namibia's Skeleton Coast. In the Gir Forest the Asiatic lion preys on sambhar, axis deer, nilghai, wild boar, and also domesticated livestock, especially in times of drought.

Siberian Tiger (*Panthera tigris altaica*)

This is the largest cat, a massive animal with heavily muscled shoulders and forelimbs. The average weight of an adult male is about 650 pounds (295 kg) with a record of 840 pounds (381 kg), and such large animals would measure 12 feet (3.6 m) from nose to tail tip. The Siberian tiger is the subspecies adapted for life in the harsh winter conditions of eastern Russia, where the temperature may reach −40°F (−40°C). It has a long yellowish coat in winter, which becomes reddish in summer, with a creamy-white belly and flanks. There are white patches over the eyes and the muzzle is white, the ears are black with a white spot on the outside and white inside, and the tail is black and white. The Siberian tiger's coat is much thicker and longer than the other races of tiger; it has less stripes than the others and they are more brownish than black, and there is little striping on the outside of its front legs. It is now a very rare animal due to poaching and the loss of its forest habitat from clear-cutting and fires. Originally occurring from Lake Baikal to Russia's Pacific coast, it is now confined to the heavily forested Ussuri Region of eastern Russia in the Sikhote Alin Mountain Range, and possibly in North Korea. The wild population is about 350 animals, but it is well represented in zoological gardens. The tiger's natural prey includes red deer, sika, and wild boar, and it is claimed that an adult male can eat up to 88 pounds (40 kg) of flesh per day, covering the remainder of its kill with grass and leaves for later use.

Siberian Tiger   *The largest cat, the Siberian tiger is now a very rare animal, restricted to a small area of forest in far-eastern Russia and neighboring North Korea. It hunts red deer, sika deer, and wild boar, mainly at night, caching for later what it cannot eat.*
Photo: Courtesy Harcourt Index

■ WILD DOGS

The animals we call foxes, wolves, dingoes, coyotes, and several other names are all members of the dog family *Canidae*, which is divided into two groups—the dog-like species and the fox-like species. Zoologists know them as canids, but it is acceptable to call them all "dogs." The wild dogs are distinctive animals, their form generally long and supple, with long legs, a heavy chest, usually a bushy tail, erect ears, a long muzzle, and nonretractile claws. They vary in size from the tiny fennec fox to the northern wolves (although domestication has produced both smaller and larger breeds), and are mainly of uniform coloring, the major exceptions being the striped jackals and the blotched African hunting dogs. They employ two basic methods of prey capture, either by the listening, stalk, and pounce characteristic of the coyote and red fox; or the long chase of the strategic pack hunters such as the dhole and wolf, which rely upon exhausting their quarry and are able to bring down large animals through their combined efforts. They do not waste energy on a long chase of small prey, as the effort expended would not be adequately returned in flesh and energy. However, although dogs are hunters, they are not as strictly carnivorous as the cats, and in fact are really omnivores. Their natural intake includes a high percentage of carbohydrates such as fruit, berries, cereals, and vegetables, all of which they can metabolize, unlike the cats, and they also eat herbivores' droppings, especially those of rabbits, when B vitamins are lacking in their diet. The dog's carnassial[5] teeth have changed slightly and have a grinding edge behind the cutting edge, allowing the chewing of a more varied, omnivorous diet. Several nocturnal species, such as the raccoon-like dog and the jackals, are quite omnivorous and should perhaps be included with similar animals at the end of this chapter, but they are hunters, even though they may eat non-animal foods when the opportunity arises.

Smell is the most important of the dog's senses, beginning at birth when the mother licks her pups and places her scent on them and then licks her teats, for the blind pups may otherwise have difficulty finding them. The pack hunters discover their prey mostly by smell, rely upon it initially during the chase, and then depend heavily upon sight as they get close to their quarry. The sense of smell is also important to the solitary hunters such as the red fox and coyote, particularly for males to locate receptive females, and with their exceptional hearing they can locate mice in the grass or beneath the snow.

The degree of smell in mammals depends upon the size of the nasal epithelium, and the larger the epithelium the more receptors it can contain and the greater the sense of smell. Humans have a small epithelial surface and about 5 million receptor cells, whereas dogs have a much larger area and about 250 million olfactory receptor cells. Dogs have been so intent on tracking with their muzzle to the ground that they have missed other prey in clear view. Domesticated dogs are able to detect variations in human sweat due to illness or emotion, and a highly developed sense of smell is essential for communicating, as scent-marking with feces and urine are an important aspect of a dog's life. Dogs also have a vomeronasal organ in the roof of the mouth to process sex-scents and send signals to the brain.

Their sense of taste works in conjunction with their sense of smell, and dogs have about 1,700 taste buds in their mouths, more than the cat which has 475, but far less than the 9,000 buds in the human mouth. These are located on the tongue, palate, and epiglottis, and in domesticated dogs they are known to test the palatability of food, and to stimulate salivary and gastric secretions. Dogs are very sensitive to bitter tastes and can appreciate sweet tastes.

The dog's night vision is superior to a human's, and like the cats they have a tapetum which reflects light back through the retina, and produces "night shine." They have widely spaced eyes for good lateral vision, but poor binocular vision as the fields of view of each eye barely overlap, so less of what they see is in focus. This gives them a wider angle of view than the cats, and they are good at seeing peripheral movement but they cannot focus on objects closer than 12 inches (30 cm). They have partial color vision, being able to see greens and blues.

The sense of hearing in dogs ranges from 67 Hz to 45 kHz, more than double the human range at the higher end, which is why shepherds can control their dogs with whistles that we cannot hear. They can move their ears independently, using one ear to scan for sound and both to capture and funnel sound waves, and can then locate the source of a sound in about half a second. Their well-developed hearing is illustrated by coyotes, which can locate mice by their tiny rustling noises in the grass or beneath the snow. The outer ears of all the carnivores reach their greatest development in the foxes, especially the bat-eared fox (*Otocyon megalotis*), and the fennec fox (*Fennecus zerda*). Dogs have touch receptors all over their bodies, especially on the feet, and they have vibrissae, which sense air flow, above the eyes and on the muzzle and chin. The wild dog's basic senses have been capitalized upon by breeders of domesticated dogs, using hounds such as the Afghan and saluki for hunting by sight; beagles for sniffing out explosives, drugs, and even fruit at airports; and bloodhounds, whose sense of smell is so effective and accurate that its "evidence" in following a human trail is accepted in a court of law.

Some of the Species

Fennec Fox (*Vulpes zerda*)

The fennec fox is without doubt the most charming of all the small dogs, with a soft coat of pale sandy color and a bushy, black-tipped tail. It is also the smallest member of the wild dog family, with a head and body length of 16 inches (40 cm) and a weight of 3¼ pounds (1.4 kg). Hearing is its most highly developed sense for it has the largest ears in proportion to its size of all the wild dogs, able to pick up the smallest sounds while also helping to radiate heat from the body. It also has large bullae or earbones which improve hearing sensitivity. The fennec fox lives in the desert zones of North Africa and the Middle East, where it preys upon gerbils, jerboas, insects, and lizards, and is also fond of dates. It gets sufficient fluids from its food to manage without free water for long periods. A burrower, it digs tunnels several yards long where it spends the daylight hours, appearing at dusk to hunt, when its hairy soles allow it to run fast on soft sand. It is now being bred commercially as one of the latest "alternate" house pets.

Red Fox (*Vulpes vulpes*)

An almost cosmopolitan animal, the red fox is probably the most widely distributed carnivore, living throughout the Holarctic Zoogeographic Region which encompasses the whole of Eurasia (excluding Southeast Asia), North Africa, and North America, yet across this vast range is considered to be a single species. Its coat is an unmistakable reddish-brown, with white belly and black legs and feet, plus the distinctive reddish-black bushy tail with a white tip (see color insert). The red fox favors both forest and open country and is a very versatile, adaptable animal with a reputation for intelligence and cunning. Although it has traditionally preferred areas of mixed habitat, such as regions where woodland, grassland, and agricultural land combine, it has also become quite urbanized in some areas, living completely within cities, raiding garbage dumps and hunting rodents in city parks. It is killed for sport, to control rabies, and for its pelt, and there is still a thriving fox farm industry.

The red fox is mainly nocturnal and crepuscular, although it may occasionally be seen during the day, and its senses of smell, hearing, and vision are all well developed. Family groups have a main denning area, which includes burrows up to 30 feet (13.5 m) long and 6 feet (1.8 m) deep, and several alternate emergency burrows elsewhere within their territory. The main den may be used for many years by succeeding generations. The red fox eats almost anything edible, including rodents, rabbits, hares, deer fawns, pheasants, partridges, insects, snails, berries, fruit, cereal crops, and fungi.

Black-backed Jackal (*Canis mesomelas*)

A slender, coyote-like dog with large ears, the black-backed jackal exists in two separate populations, in east Africa and southern Africa. It weighs about 30 pounds (13.6 kg) and is easily identified by the black saddle that extends from its nape to its rump. Jackals are animals of the dry grasslands, scrub country, open woodland, and the forest edges. They vary in their activity patterns, often depending upon the presence of people, and they may be nocturnal, crepuscular, or diurnal. They are normally social and territorial animals, occurring in small groups comprising a mated pair and their growing young, and often with older female offspring, which assist in raising litters but do not themselves breed. Although jackals scavenge from garbage dumps and the kills of other carnivores, they are not total scavengers as once believed, and their diet also includes animals they hunt and kill themselves, such as young antelope, rodents, hares, guineafowl, and spurfowl. They also eat the eggs and young of ground-nesting birds, and are predators of small domesticated livestock. Family parties hunt as a unit to attack small antelope such as duikers and dikdiks and the calves of larger antelopes, just as the spotted hyenas cooperate to overcome larger prey. They are even more omnivorous than most dogs; they readily eat plant matter, including roots, berries, and wild melons, and have raided native gardens for vegetables and corn.

Coyote (*Canis latrans*)

Often mistaken for the wolf, with which they have hybridized in the wild, coyotes are smaller and have a more pointed and narrower muzzle and taller ears. Their long coats are usually buffy-gray with paler underparts, and some black shading on the shoulders, rump, and the tip of the bushy tail. Males weigh up to 39 pounds (17.7 kg) and females about 33 pounds (15 kg). Their range includes a wide variety of habitat in North and Central America—open forest, rocky scrub, semidesert, and grasslands—and like the red fox they have invaded suburbia. Coyotes are opportunists; they catch their own small prey, usually rabbits and rodents which they stalk and pounce on, and occasionally larger prey such as white-tailed deer and elk fawns. They also eat fish and frogs, the eggs and young of ground-nesting birds, and they scavenge on garbage dumps and from the kills of larger predators; but they rarely eat plant matter. They have traditionally been blamed for considerable livestock predation, especially of sheep; and after the wolf they are the most controversial predator in North America regarding their control for livestock protection and preventing the spread of rabies.

Dingo (*Canis domesticus*)

The dingo is Australia's wild dog, a placental carnivore in a land of marsupials, but it did not evolve there or arrive naturally. It is a descendant of domesticated dogs that were introduced into Australia thousands of years ago by the early aboriginal colonists. It is therefore in reality a feral animal, a domesticated breed that has reverted to the wild. Nocturnal, crespuscular, and occasionally diurnal, the dingo is now threatened by shooting and poisoning, and as a result of hybridizing with domesticated dogs. It is a medium-sized dog about 36 inches (92 cm) long, it stands 24 inches (60 cm) tall and weighs up to 41 pounds (19 kg). Normally sandy-yellow, there is a great deal of variation in color, including black animals, creamy ones, and even black-and-tan individuals, all normally having a pale belly and throat. It occurs in a wide range of habitat, especially the dry plains and scrublands of the outback, and tropical coastal beaches and alpine forest. It varies in its social habits, usually being solitary, but also roaming in small family packs. Its main food items are kangaroos and wallabies, rabbits, lizards, and birds, and it also scavenges, but its liking for sheep resulted in the construction of the "dingo fence"—a 3,000-mile (4,825 km) barrier across Australia, from the Great Australian Bight to the coast of southern Queensland, to keep them out of sheep country—the southeastern corner of the continent.

Dingo *The wild dog of Australia—actually a feral domesticated animal—was introduced there long ago by the first aboriginal colonists from Melanesia. It hunts mainly at night, and kangaroos and wallabies are its main prey.*
Photo: Courtesy Harcourt Index

Wolf (*Canis lupus*)

Although wolves are the largest wild dogs, there is considerable size variation and many recognized subspecies within their vast range across the Northern Hemisphere, one of the widest distributions of all terrestrial mammals. In the

North American Arctic regions male wolves have attained a head and body length of 5 feet (1.5 m) and a weight of 165 pounds (75 kg), and females 120 pounds (55 kg). Wolves at the southern end of the species range may be only half the weight of the Arctic animals, while animals in the Middle East may weigh only 44 pounds (20 kg).

Wolves also vary in color, the most frequent being grayish-brown with black shading on the chest, shoulders, and tail, and pale buffy-white legs and underparts; but melanistic and albino mutations occur frequently, the latter especially in the Arctic. They are animals of many habitats including dry grassland, rocky scrub, forests, and tundra. Wolves are usually considered the most social wild dogs, living and hunting in packs of about ten animals—normally comprising a mated pair and their offspring. The large pack size allows the hunting of large prey, such as musk oxen, bison,[6] moose, elk, and caribou in North America, and wild boar, nilgai, red deer, and reindeer in Eurasia. All these ungulates are fair game to wolves, which cover long distances in search of prey and may then chase their victim for several miles, finally getting close enough to snap at its rump and belly until it succumbs. They also scavenge animals that die naturally and usurp the kills of other carnivores. The wolf has been eliminated from most of the continental United States, but still occurs in Minnesota, Michigan, and Wisconsin, plus the northern Rocky Mountain states, and has been reintroduced into Yellowstone National Park, and Mississippi, South Carolina, and Florida. It is common in Canada and Alaska, rare in Europe, and even more so in North Africa; it is becoming scarce in southwestern Asia from Pakistan to Nepal due to loss of habitat and reduction of its prey, but especially to persecution for real and suspected livestock predation, and the long-perceived but unsubstantiated threat to humans.

■ WEASELS AND THEIR RELATIVES

The weasels, ferrets, mink, and their relatives belong to the family *Mustelidae*, the second-largest[7] family of carnivores, containing sixty-five species that vary considerably in their appearance and lifestyle. This very mixed group of small mammals contains nocturnal and diurnal species, sleek, fast-running pursuit hunters, and others that shuffle along slowly with their nose to the ground. There are arboreal forms, squat-bodied burrowers, and lithe swimmers, and they have a wide distribution around the world. They are at home in northern coniferous forests and tropical rain forests, in cold northern seas and in Amazonian rivers, and the nocturnal members of the family include both hunters and omnivorous species, but carnivory predominates in the family.

These carnivores, which are collectively called mustelids, walk either on their toes (digitigrade) or they combine walking on their toes and soles (plantigrade), and they hunt mainly by scent, although hearing and vision are also well developed. Three-quarters of the mustelids are active at night, including the weasel, mink, polecat, stoat, and the black-footed ferret, which are all predators; and the honey-badger, badger, and skunk, which are omnivores. The predatory mustelids are aggressive, lithe, and quick hunters, which actively chase their prey; the weasels and polecats follow rabbits down their holes and the martens are as agile as the squirrels they chase through the trees. In contrast the ratel just uses brute force and powerful jaws to overcome porcupines, snakes, large antelope calves, and even other predators such as foxes and jackals.

Some of the Species

American Mink (*Mustela vison*)

A large weasel that weighs up to 6½ pounds (3 kg), with a natural range in North America from Alaska and Labrador to the Gulf of Mexico, the mink is now established as a result of escapes and deliberate introductions in Iceland, Great Britain, Europe, and Soviet Asia, where the native European mink is now very rare due to the competition. The American mink is a terrestrial animal that prefers wetlands—marshes, muskeg, riverbanks, and lakesides. It burrows just above water level into river banks and also uses beaver lodges and muskrat houses, and dens among rocks and under tree roots, where it makes a nest of grass and feathers. It is almost totally carnivorous, quick enough to catch fish, but relies more on amphibians, rodents, and aquatic birds, and it may eat berries in the autumn. Its

Black-footed Ferrets   *Nocturnal members of the Mustelidae family, black-footed ferrets were dependent upon prairie dogs and suffered as they were eradicated. The last wild specimens were caught in 1987, but captive-breeding programs by the U.S. Fish and Wildlife Service have been successful and allowed their reintroduction into the Wyoming prairie.*
Photo: Courtesy USFWS, LuRay Parker

soft and luxurious pelt has been coveted by the fur trade for many years, and the great commercial demand led to large-scale trapping and eventually to fur farming, which has resulted in its domestication. The mink's natural color is rich blackish-brown above, with paler underparts, but domestication has produced several color mutations. It is a very agile and versatile animal; it climbs and swims and is a good jumper. Mainly nocturnal, solitary, and territorial, it is very aggressive to conspecifics except at breeding time.

Polecat and Ferret (*Mustela putorius*)

The ferret is the domesticated descendant of the polecat, but its origins are lost in antiquity. It has been used for centuries in England and Europe for rabbiting, which seems the most likely region and reason for the initial domestication of the polecat. It has been suggested that the ancient Egyptians were the first to domesticate polecats for "hunting," which seems rather unlikely as the polecat's natural range does not extend to the Middle East or North Africa, and it is unclear what animals they could possibly be trained to hunt there, although they were used to control vermin.

Polecats are nocturnal animals and the domesticated ferrets have a tendancy to follow suit, but with regular handling from infancy they become more diurnal. In the wild the polecat is a seasonal breeder that mates in spring, but as a result of man's control, ferrets now breed throughout the year. They have become a favorite house pet in recent years, available in two color phases, creamy-white and "polecat," which is a brownish animal resembling its wild progenitor.

The wild polecat has a head and body length of 18 inches (45 cm), its tail is 7 inches (18 cm) long, it weighs 3 pounds 8 ounces (1.6 kg), and it is slightly larger than the domesticated ferret. It has pale-yellow underfur with blackish guard hairs, off-white ear tips and face, and a broad dark mask across the eyes. The polecat is a native of the British Isles, Europe, and western Asia, where it lives in forests, grasslands, and on mountain slopes; but both it and the ferret have been introduced and are established in New Zealand. The polecat is endowed with an extraordinary sense of smell, and is sensitive to sounds ranging from 16Hz to 44,000Hz. Its main prey are rabbits, rats, amphibians, ground-nesting birds such as grouse and pheasants, and it raids poultry pens. In turn it is preyed upon by wolves, lynx, foxes, and eagle owls.

Wolverine (*Gulo gulo*)

Wolverines are the largest land mustelids, with adult males reaching a weight of 66 pounds (30 kg) and with a head and body length of 39 inches (1 m), and a short but long-haired tail of 10 inches (25 cm). Females are only two-thirds the size of the males. They have stocky bodies, short and sturdy legs and large feet, a large head, and small ears. Their long coats are dense blackish-brown, with a pale-brown band along the sides of the body from the shoulders to the tail and joining over the rump.

Wolverines are pan-Arctic animals, of the northern tundra and the taiga, ranging south to California and Pennsylvania in North America and Germany and Siberia in Eurasia. Mostly terrestrial, they are the triathletes of the animal world, excelling at swimming, running, and climbing. They have great stamina, can lope steadily nonstop for up to 9 miles (15 km), and are active both day and night. Powerful, tenacious, and fearless hunters, they have driven cougars and sometimes bears from their kills, but they in turn have been driven off their kills by packs of wolves. The wolverines' large feet are an advantage on soft snow when chasing hoofed mammals, and in addition to eating carrion they hunt deer, reindeer, rabbits, hares, and grouse, and are well known for their habit of raiding remote cabins and food caches. Their sense of smell is highly developed, but their sight and hearing are poor.

Ratel or Honey Badger (*Mellivora capensis*)

The Old World's equivalent of the wolverine, the ratel is an incredibly strong and fearless animal, similar in size to the Eurasian badger. It has a wide range from the Middle East eastward to India and southward to South Africa, but is now very scarce everywhere except in Africa. It is a stocky animal with short legs and tail, a massive head with powerful jaws and strong teeth, and small ears and a blunt muzzle. The ratel's coloration is the reverse of the normal mammalian scheme in which the upperparts are the darkest and the belly pale, for it has whitish-gray upperparts and is black underneath, which is warning coloration that it is an animal to avoid. It is absolutely fearless, and attacks much larger animals in defense of its den or young, plus creatures that are no threat, such as passing buffalo, injuring them severely. It digs fast and runs fast and tirelessly in its preferred habitat, which is grasslands, bush, and wooded savannah. The ratel has a rather famous association with a bird called the honey guide and honey bees, in which the bird supposedly attracts the ratel with its calls to a bees' nest and then feeds on the bee larvae when the ratel rips the hive open to get the honey. It certainly destroys man-made beehives and raids poultry pens, and is a hunter of young antelopes, rodents, hares, reptiles, and ground-nesting birds. It also eats carrion and a wide variety of plant matter including fruit, berries, tubers, and wild melons, and so it is an omnivore, but there is absolutely no doubt that it is a predator.

■ HYENAS

There are four species in the family *Hyaenidae*, but despite their similarity to dogs, they are taxonomically closer to the *Viverridae*—the civets, genets, and mongooses. Three species are hunters and scavengers, while the fourth—the aardwolf—is primarily an insect-eater, and although technically carnivorous as it is an eater of animals, it is certainly not a hunter.

The hyenas have a characteristic shape: a stocky body with a large head and shoulders and sturdy legs that are longer in front, resulting in a pronounced slope from the nape to their poorly developed hindquarters. They have blunt, dog-like nonretractile claws, and anal scent glands with storage pouches. The family is restricted to Africa and southern Asia as far east as Bangladesh, and they are all animals of the grasslands and open woodland that are mainly nocturnal in their activities, spending the daylight hours in caves, in the abandoned holes of other animals, and occasionally in tunnels of their own digging.

The spotted hyena is the most social species, living in packs with the members cooperating to attack large herbivores. The brown hyena and the striped hyena are less social and live a more solitary lifestyle, relying more on scavenging and on smaller prey which they can overcome themselves. Striped hyenas have banded together to attack domesticated animals, and have killed people. The brown hyena, while known to be the least predatory of the three, is regularly accused of killing domestic livestock in Namibia and Botswana, and it kills fur seal pups on the Namib Desert coast.

All three species have short, powerful jaws, and the spotted hyena's skull is specialized for crushing bone and cutting hide, with large, bone-crushing pre-molars and shearing carnassial teeth. The others have relatively small molars and their carnassials crush as well as shear, indicating a more omnivorous diet, or at least one less dependent on scavenging from large mammal carcasses. The hyenas have well-developed senses of hearing, sight, and smell, and they use the secretions of their anal glands to mark their territorial boundaries.

Some of the Species

Spotted Hyena (*Crocuta crocuta*)

The spotted hyena is the largest and most powerful species, a solid-looking animal that weighs up to 176 pounds (80 kg) and has a shoulder height of 36 inches (90 cm). The characteristic hyena shape is most obvious in this species, with its large and powerful head, neck, and shoulders, sloping down to the weak hindquarters, with a short, tufted tail and shorter ears than the other species; it is yellowish-gray with dark spots over the whole body and upper legs. Sight, hearing, and smell are all well developed in this species, which in proportion to its size has the most powerful jaws of all animals, capable of crushing the largest bones to reach their marrow. Originally believed to just scavenge, it is now known to be an efficient hunter that probably gets more of its food from killing than scavenging, and lions may more often take advantage of hyena kills than the reverse. It runs very fast over short distances, reaching a speed of 36 mph (60 kph), and can run down its victim, particularly antelope calves and zebra foals, and animals discerned to be weak through age or sickness.

The spotted hyena is nocturnal and dens by day in burrows of its own digging or those usurped from porcupines, aardvarks, and warthogs. Its historical range in sub-Saharan Africa has been reduced by loss of habitat, persecution as a predator of livestock, and as a threat to humans sleeping in tents and doorless native huts; and it is now rare in many parts of its range, particularly in southern Africa.

Spotted Hyenas *A pack of spotted hyenas make short work of their prey on the East African plains. Once believed to be just scavengers of lions' leftovers, they are now known to band together to actively hunt antelope, zebras, and gnu, and lions may actually scavenge their kills.*
Photo: Photos.com

Striped Hyena (*Hyaena hyaena*)

This is the hyena with the widest distribution—from Tanzania and Senegal through North Africa, Asia Minor, and across southern Asia to Bangladesh. It is an animal of the dry grasslands and rocky scrub, ranging from sea level up to 9,850 feet (3,000 m). Smaller than the spotted hyena, the sexes are similar in size, standing about 33 inches (84 cm) at the shoulder and weighing up to 110 pounds (50 kg). The striped hyena's ground color is grayish-brown, heavily striped with dark brown, vertically on the body and horizontally on its slender legs. It has a long, dark erectile mane on the spine and a longer and bushier tail, about 18 inches (47 cm) long, and much longer ears than the spotted hyena. It is also nocturnal and sleeps during the day in abandoned burrows or holes of its own digging.

In East Africa the striped hyena is mostly unsocial and a scavenger, but in North Africa and across southern Asia, where large game is now scarce, it lives in family groups that cooperate to hunt wild boar and wild ass foals, and also attack domesticated donkeys, sheep, and goats. It kills small animals such as hares, porcupines, tortoises, lizards, and birds, and like the brown hyena it is a seashore scavenger; it also eats melons, berries, and cultivated cereal and vegetable crops. Its status varies from being locally common to quite rare, especially in northwestern Africa.

■ GENETS AND MONGOOSES

The largest family of carnivores is the *Viverridae*, which contains seventy-one species of small mammals including the familiar mongooses and the less-well-known genets, civets, and palm civets. Some taxonomic authorities now place the mongooses in a separate family, the *Herpestidae*. These are animals of the Old World only, occurring in southern Asia, Indonesia, and Africa, with just one species in southwestern Europe. They vary widely in their diet, with several species being active predators, whereas others are content to eat fruit, plus any insects, worms, snails, birds' eggs, or helpless small animals such as nestling birds or baby rodents which they come across on their rambles. Most of the members of this family are decidedly nocturnal, and only in the mongoose subfamilies *Galidinae* and *Herpestinae* are most of the species active by day.

The *Viverridae* is currently divided into the following subfamilies:

Viverrinae—18 species of civets, genets, and linsangs, all of which are nocturnal, and the genets and linsangs are hunters.

Paradoxurinae—8 species of palm civets, including the binturong, all nocturnal omnivores.

Hemigalinae—4 species of banded palm civets and related forms, all nocturnal omnivores.

Fossinae—the Malagasy civet and falanouc, which are both nocturnal omnivores.

Galidinae—5 species of Madagascan mongooses, of which 2 species are nocturnal hunters.

Herpestinae—33 species of mongooses, of which only 5 are mainly nocturnal hunters.

Cryptoproctinae—a single species, the fossa, which is a nocturnal hunter.

Some of the Species

Large-spotted or Blotched Genet (*Genetta tigrina*)

Genets are attractive, cat-like animals with long and slender bodies, about 20 inches (51 cm) in length, a tail almost as long, and weigh about 6 pounds (2.7 kg). The large-spotted genet lives in sub-Saharan Africa, excluding the arid, south-western zone. Its coat is short and dense, usually pale gray with many large, blackish-brown spots or blotches, and it has a black-ringed, black-tipped tail with a dark dorsal stripe from its shoulders to tail. The face is whitish with a black band across the center of the muzzle. Melanistic individuals are common, but there is considerable geographic variation in the species, with animals from the drier regions being pale with subtle markings, and those from moist forested areas being darker with more definite patterns. Although found in a wide range of habitat it prefers dense vegetation, especially moist riverine forest, and agricultural areas where it preys heavily on rodent crop pests. The genet is a solitary animal, quite nocturnal in its behavior and mostly arboreal, which sleeps during daylight high in the foliage or in tree holes. It is very agile, leaping nimbly between trees and jumping down to the ground from several meters, and its slender and loose-limbed body allows it to pass through holes that its head can enter, like the weasels. It has short and curving retractile claws and anal glands that emit a musky discharge. Genets actively hunt rodents and birds, invertebrates and reptiles, and they also eat fruit.

White-tailed Mongoose (*Ichneumia albicauda*)

This is one of the seven species of mongooses (including the two Malagasy species) that are wholly or partially nocturnal. It does not like moisture and lives in woodland, savannah, and semidesert country in sub-Saharan Africa, from Senegal to Sudan and then south to Cape Province, and also in the southern Arabian Peninsula. During the day it sleeps in aardvark and porcupine burrows, in termite mounds, or in thick undergrowth, and is rarely seen at night either, unless caught in car headlights; but it is often heard as it is a very vocal animal with a dog-like yap. The white-tailed mongoose is one of the larger species of mongooses, with long hair, long legs, a long and narrow head, and large, rounded ears. It is yellowish-brown with long black guard hairs, which give it a grizzled appearance, and its legs and arms are black. The large, bushy tail is yellowish-brown at the base and white on its terminal half. This mongoose walks and trots with its head and shoulders lower to the ground than its rump, and for defense it secretes the very obnoxious contents of its anal glands. Its diet includes invertebrates, rodents, lizards, and snakes, and it raids poultry pens and breaks eggs by throwing them backward between its legs onto a rock.

■ PART-TIME CARNIVORES

According to the fossil record the mammalian carnivores began evolving about 60 million years ago. Initially, their food preferences were obviously carnivorous, and over the course of their evolution they developed the adaptations for survival which included improved senses for prey location, increased speed and power, modified teeth and jaws for overcoming their prey, and an unspecialized stomach to digest large amounts of meat. But they did not all remain strict meat-eaters, and many have since changed to a more omnivorous diet; a few have even become vegetarians. As a result certain changes have occurred in keeping with their new way of life. The cutting or carnassial teeth of most bears and the raccoons, for example, became adapted for an omnivorous or herbivorous diet, with broad and flat premolars and molars for crushing replacing the carnassials for cutting. The gut of these species remains unspecialized, however, like the diurnal giant panda, which must digest its diet of plant fiber in a carnivore's stomach, and must eat continually all day to get sufficient nourishment from its poor-quality diet, in contrast to the herbivores in which bacteria break down the fiber into digestible components. Also, despite modifications to their cheek teeth these aberrant carnivores, even the fruit-eating kinkajou, still have their long and sharp canine teeth.

These nocturnal converts to omnivory and even to frugivory still prey on live animals. The civets catch mice, the raccoons catch crayfish and frogs, badgers dig ground squirrels from their burrows, and the skunks and hog badgers seize small snakes, lizards, and nestling birds; but none can be considered hunters.[8] They cannot match the hunting prowess of the cats, dogs, or weasels; and unlike the cats and weasels they do not rely solely on meat.

Some of the Species

Bears

Bears belong to the family *Ursidae*, a very distinctive group of mammals containing the world's largest terrestrial carnivores. They have stout bodies and large heads, short limbs, short tails, and small eyes and ears. Their mode of walking, in which the heel and sole are placed on the ground, is called plantigrade. They have five digits on each foot and long and powerful claws, and their soles are hairy around the pads, except in the sun bear which spends a lot of time climbing trees. All bears have a good sense of smell but poor sight and hearing, with the exception of the polar bear, which has very good vision. All bears except the carnivorous polar bear have wide, flat-topped molars for grinding food, as a result of their omnivorous diet, but although they are no longer regular hunters in the carnivore manner, they do occasionally actively hunt other animals. Eight species are recognized but only the sun bear and sloth bear are nocturnal, and they are both omnivorous.

Sun Bear (*Helarctos malayanus*)

The sun bear is the smallest bear, which weighs up to 132 pounds (60 kg) and has a shoulder height of only 28 inches (70 cm), but despite its size it has very powerful limbs and claws, and walks with a peculiar gait, with all four feet turned inward. Its coat is short and shiny, blackish-brown in color with a gray muzzle, and usually has a pale-orange patch on the breast which varies in shape and size. It is a very solitary creature, and if two are seen together they are invariably a mother and her cub. The sun bear is the most nocturnal bear, a native of the equatorial forests of mainland Southeast Asia, Sumatra, and Borneo, where it is quite arboreal and usually sleeps by day in a tree on a bed of broken branches. It is also the most tropical bear, and with food available all year it does not need to hibernate. An omnivore, it eats whatever it can acquire, favoring ripe fruit and honey, ripping open trees with its powerful claws to access wild bee hives, and it damages coconut plantations by tearing open the young palms to eat their centers. Its animal food includes carrion, rodents, and the eggs and young of ground birds such as pheasants and junglefowl. In temperament it is a shy, cautious, and very moody and aggressive animal.

Sloth Bear (*Melursus ursinus*)

This is an unmistakable bear, with a long, shaggy coat and a neck ruff, and a long and very broad snout. It is usually black, although dark-brown or reddish-brown individuals are occasionally seen, and has a large U- or V-shaped, whitish-yellow mark on the chest. It is a medium-sized bear that weighs up to 320 pounds (145 kg) and stands 31 inches (79 cm) at the shoulder when adult. The sloth bear is a tropical animal, found in India, Sri Lanka, and Bangladesh and in the Himalayan foothills of Nepal. Although it is a forest-dweller it rarely climbs, and prefers the more open and dry forests with rocky outcrops, where ants and termites are plentiful. These form a large part of its diet and are located by smell, as it has poor vision and hearing; it has been likened to a vacuum cleaner for the manner in which it gathers termites. The modifications for this lifestyle include a gap in its front teeth—through the loss of the inner upper incisors—and a hollowed palate so that its mouth forms a funnel. It blows dirt off the termites and then inhales them with a noisy sucking action that can be heard some distance away. It is now very much reduced due to loss of habitat, human persecution for crop damage and occasional livestock predation, and its danger to humans, which is justified, although this usually results from accidental contact often due to the bear's shortsightedness and poor hearing. It is nocturnal and sleeps in dense bush or among rocks during the day.

Aardwolf

Like the hyenas, the aardwolf (*Proteles cristatus*) is a member of the family *Hyaenidae*, but is the only species within its genus, a unique animal in both appearance and diet, for it eats mainly termites. It is actually a degenerate, aberrant, or atypical hyena, which physically resembles a cross between a dog and a hyena, and is reminiscent of a small striped hyena. The changes to its behavior and form have resulted from taking advantage of a very plentiful source of food and adopting a primarily insectivorous diet. It is believed to have been a fairly recent convert to such a diet, however, with subsequent weakening of its jaws and teeth. It is not equipped, like the aardvark, to break into termitaria, and therefore feeds almost exclusively on *Trinervitermes* termites, which live in mounds but forage outside at night for grass; this is where the aardwolf licks them up. It also eats other insects and the maggots it finds on decomposing carrion, and possibly also eats some carrion. As a result of its diet, the aardwolf has lost the typical powerful skull and jaws of the scavenging hyenas, and its formerly carnivorous dentition—the huge premolar and molar teeth—has degenerated to small, widely separated stumps; but its canines are long and sharp and have likely been kept for defense. It has a humpbacked appearance, with sloping lower hindquarters, long legs, long neck, and a pointed muzzle. Its long and soft coat has a thick underfur and coarse guard hairs, and is longer along the spine; its 12-inch-(30 cm)-long tail is quite bushy. In color the aardwolf is yellowish-gray with black stripes, and is quite small compared to the other hyenas, weighing only about 26 pounds (12 kg) and standing just 20 inches (50 cm) at the shoulder.

Aardwolves live in the grasslands and dry scrub of the eastern half of Africa from northern Sudan to Cape Province. They are strictly nocturnal, mainly solitary and territorial animals, which normally den close together; but they have been observed in family groups and in bands of adult females that have denned together to raise their young. Like the real hyenas they mark their territory with their anal scent glands.

Badgers and Skunks

Badgers and skunks are the nocturnal and omnivorous members of the families *Mustelidae* and *Mephitidae*, respectively, all of which eat a wide range of foods of plant and animal origin. The badger family contains the hog badger, stink badgers, ferret badgers, and the Eurasian and American badgers, whereas there are nine species of skunks within the *Mephitidae*. The badgers are mainly terrestrial mammals, which dig their own burrows with their powerful forelimbs and long claws, and also dig out much of their food, especially tubers and ground squirrels. Only the ferret badgers are able to climb trees, where they often prefer to sleep rather than underground like the others. All the badgers have anal scent glands, and the potent secretions of the stink badgers are almost on par with a skunk's discharge.

The Eurasian and American badgers are very similar, differing mainly in size and their striking black-and-white head markings. They are mainly northern animals, one distributed across northern Europe and Asia, the other in North America, and both practice delayed implantation, with mating occurring throughout the summer; but the fertilized eggs remain suspended as blastocycts in the womb and do not begin their development until late December. Following a "normal gestation" period of about seven weeks birth then occurs in February. Both species hibernate for weeks or months in midwinter, but the tropical forms—the hog badger, stink badgers, and ferret badgers—do not need to hibernate. Skunks are instantly recognizable animals, and their capabilities are very well known, for their secretions are the most noxious of all animals. Their bold black-and-white coloring makes them visible and memorable to potential predators, which are loathe to repeat their experience. The skunk's musk glands lie next to the anus, and have two outlets from which the scent is expelled by muscular action as a fine mist or larger droplets, and its pungency and durability deserve their reputation.

The skunks are restricted to the New World. In North America the little spotted skunk and the striped skunk occur in southern Canada and then south throughout the United States. The other species have a more southerly distribution, in the southern United States and in Central and South America. They den in their shallow burrows or in hollow logs during daylight and when they appear at dusk they are themselves preyed upon by owls, although road traffic is now their major threat in many areas. Delayed implantation also occurs in the northern skunks but for a shorter period than the badgers. The skunks eat a wide range of food, including insects, earthworms, rats and mice, birds, eggs, snakes, berries, fallen fruit, acorns, and vegetable crops such as tomatoes, beans, peas, and peanuts.

Little Spotted Skunk (*Spilogale putorius*)

This is the smallest of the skunks, just 12 inches (30 cm) long and weighing 2 pounds (1.8 kg). It has typical warning skunk coloration, black with white markings, the patterning being so variable that each animal is different, but basically it is black with white stripes on the back and sides, which break up into spots and stripes on the rump, and it has a white patch on the forehead. It has short hairs on its face, longer hairs on the body, and a white plumed tail almost 8 inches (20 cm) long.

The little spotted skunk has the typical skunk scent glands and raises its tail as warning of intent; if this is ignored it then stands on its forefeet and raises its hind legs into the air, dropping down again to discharge the glands. It has a wide range, from southern Canada throughout most of the United States into northern Mexico, and hides during the day beneath buildings, in hollow logs, or in burrows borrowed from other animals and possibly enlarged. It prefers dry, wooded, and rocky habitat and avoids heavy forest and wetlands. Its diet is typical of an omnivore, including rodents, invertebrates, carrion, fruit and berries, and it raids chicken coops for birds and eggs. It is a vector of rabies, and even though it is believed that there are many skunks kept as pets in the United States, there is no rabies vaccine available for them yet, although some are being developed and tested.

Striped Skunk (*Mephitis mephitis*)

The striped skunk is predominantly black, with a white nape that divides into a V and extends back to the base of its bushy black tail, and when adult it may weigh 14 pounds (6.5 kg). It ranges from southern Canada to northern Mexico, in a variety of habitat, but preferring woodland, bush, and grassland, generally not too far from water, and is quite nocturnal and seldom seen during daylight. Northern skunks may stay in their winter dens for up to five months during which they survive on stored fat, but their temperature does not drop significantly so their state of torpor is therefore similar to that of the other hibernating but light-sleeping carnivores— bears, badgers, and raccoons. Dens are made under the protection of piles of brush, logs, or rocks, or beneath buildings, and several females may den together for the winter, but males are usually solitary. The striped skunk's diet includes frogs, salamanders, insects, snakes, birds' eggs and nestlings, small rodents, carrion, and garbage. It has an excellent sense of smell and hearing, but poor sight, being very nearsighted and unable to focus clearly on objects more than 10 feet (3 m) away.

American Badger (*Taxidea taxus*)

The American badger has a grayish body and a dark face, with a white line through the eyes, a white patch below the ears, and a thin white stripe from nose to nape. It lives in central and western North America from northern Alberta to

Mexico. Slightly smaller than the European badger and weighing 26 pounds (12 kg), it has the same stocky body and short legs, with the typical "flattened" appearance. Badgers live socially for most of the year in a complex system of burrows and chambers known as setts, which they dig with their powerful front feet, and in which they make nests of leaves and grass where they spend the day. They appear at dusk to hunt for young rabbits and rodents, especially ground squirrels, which they dig out of their burrows, plus reptiles, amphibians, carrion, and invertebrates (bee and wasp larvae, snails, and worms) that they locate with their well-developed senses of hearing and smell.

They also eat a wide range of plant matter including nuts, fruit, acorns, tubers, and mushrooms, and they raid cultivated crops of grain, vegetables, and fruit. Badgers are noted for their playfulness and their cleanliness—they dig holes for their feces and change their bedding regularly. They stay in their burrows when it is very cold or if there is deep snow, and may be inactive for several months in the north in a state of light torpor, living off their fat reserves.

Hog Badger (*Arctonyx collaris*)

The hog badger is a stocky and powerful animal that resembles a European badger in size and appearance, with a grayish-yellow coat, and a white face and ears with two black stripes from the nose over the eyes and ears to the neck. Adults weigh about 25 pounds (11.3 kg) and are 24 inches (61 cm) long. The hog badger has very thick and loose skin, a short white tail, and small eyes that are set widely apart; sight is not a major sense. Its head is elongated with a hairless, pig-like snout, which is very mobile and is used to root in the soil for tubers, bulbs, and invertebrates, which it finds with its keen sense of smell. Despite having such a seemingly innocuous, totally omnivorous diet, the hog badger is a savage animal with powerful jaws, strong teeth, and long claws. In addition, it can eject an obnoxious secretion from its paired anal glands. It is highly nocturnal, and shelters by day in burrows of its own digging or in rock crevices. The hog badger has an unusual discontinuous distribution, occurring in Southeast Asia from central China south to Thailand and east to Sikkim, and then on the island of Sumatra, but not in the intervening Malay Peninsula.

Raccoons and Their Relatives

The mammal family *Procyonidae*, whose members are generally called procyonids, is currently divided into the subfamilies *Ailurinae*—which has a single representative, the lesser or red panda of the Himalayas and China—and *Procyonae*, which contains a number of nocturnal New World species—the raccoons and their relatives the ring-tailed cat, kinkajou, and olingo.[9] These relatives of the raccoon may be close taxonomically but they look nothing like raccoons, and they are also very different in their behavior. The slender, ring-tailed cat has a pointed muzzle and a long tail; the rather broad-faced lesser or red panda has a thick and luxuriant tail, and the totally arboreal kinkajou has a strong, fully prehensile tail. In common they are all primarily forest-dwellers, they walk on the soles of their feet, a style known as plantigrade; they have small furred ears and their tails are quite long in all species except the raccoon; and the tails are banded in most species.

These mammals are also quite dissimilar in their food preferences. The raccoons are omnivorous, and eat a wide range of plant and animal life, including insects, crustaceans (crabs and crayfish), amphibians, snakes and lizards, birds' eggs and nestlings; plus fruit and berries including apples, juniper and mistletoe berries, blueberries, blackberries, and raspberries. The most aberrant of these carnivores are the red or lesser panda, which is mainly a folivore or leaf-eater (mostly of bamboo); and the kinkjou and olingo, which are primarily fruit-eaters. The only really active hunter in the family is the ring-tailed cat, an omnivore, but fast enough to catch mice, and which also raids poultry houses. It is the only procyonid with carnassial teeth—the cutting teeth typical of the meat-eating carnivores. The others have small, sharp premolars and broad, flat molars in keeping with a more omnivorous diet, yet have all retained their long, pointed canines, even the fruit-eating kinkajou.

Raccoon (*Procyon lotor*)

With its black mask and ringed tail, the raccoon is the most famous and recognizable of North America's mammals. It occurs throughout the United States except for some western desert regions, and extends south into Mexico and north into southern Canada. Wild raccoons are nocturnal and are only rarely seen in daylight, but pet animals usually change their habits. They are very hardy animals, but avoid the extreme low temperatures at the northern end of their range by hibernating for several months in burrows, hollow trees, house attics, and barn lofts when their fat reserves sustain them. Their sleep is not the profound torpidity of the ground squirrels, however; like the bears and other hibernating carnivores their body temperature drops only a few degrees and they are "light sleepers" that awaken quickly. They usually sleep communally to benefit from the combined body warmth of their companions. They are heavy, strong animals, adult males weighing up to 35 pounds (16 kg), and with their powerful jaws and often irascible temperament as they mature, pets may outgrow their docility and can give a serious bite. Raccoons are good climbers with great forearm mobility and dexterous hands and are omnivorous, predators of small mammals and birds, eaters of carrion and human garbage, and raiders of gardens and fields for grain, vegetables, and fruit.

Red or Lesser Panda (*Ailurus fulgens*)

The red panda lives in the mountain forests of the eastern Himalayas, and southern China, between 4,900 feet (1,500 m) and 13,000 feet (4,000 m), and is consequently a very hardy animal, accustomed to low temperatures and heavy snowfall. Arboreal and nocturnal, it is a very good climber (see color insert); it sleeps curled up high in a tree, in a tree hollow, or a rock crevice during the day and is active at night, when it seeks the bamboo shoots, which form the bulk of its diet (like the unrelated giant panda), plus invertebrates, birds' eggs, nestlings, and roosting birds. It is usually seen in pairs or small groups, which are very territorial and mark their boundaries with their anal glands' musk secretion, so its sense of smell is obviously well developed, but it is also known to have acute hearing. The red panda has a long and luxuriant coat of dark chestnut and a bushy non-prehensile tail. Its underparts are blackish-chestnut, its face and ear fronts are white and there are dark curving lines between the eyes and jaw. When adult it weighs about 11 pounds (5 kg). When asleep high in the trees during the day it is vulnerable to attack by the large yellow-throated marten, as are its cubs in the nest in a tree hollow. Snow leopards, dholes, and Asiatic golden cats are its main concerns when it feeds in bamboo groves at night.

Kinkajou (*Potos flavus*)

An arboreal and nocturnal tropical rain forest animal, the kinkajou ranges from Mexico to central Brazil. It has a very soft and woolly coat, tawny-yellow in color with a darker face, and has a round head with small, rounded ears. It weighs up to 8¾ pounds (4 kg) and has a prehensile tail 20 inches (50 cm) long. It is a solitary and territorial animal and is almost totally frugivorous, but may eat invertebrates and possibly birds' eggs and nestlings. It is very fond of honey and was once imported in large numbers for the pet trade under the name "honey bear." The kinkajou is a very fastidious eater, able to peel a grape with its teeth, and has a very long, extensible tongue as an adaptation for eating soft fruit and nectar. It is completely nocturnal and almost completely arboreal, and spends the day in a tree hollow where it is generally safe from predation, except perhaps from the diurnal tree-weasel called the tayra (*Eira barbara*). At night when searching in the forest canopy for food, it is vulnerable to the nocturnal tree-climbing cats such as the jaguar, ocelot, and margay, and to large owls and constrictors. The kinkajou has well-developed senses of smell and vision, and uses scent-marking for communicating with conspecifics. The very similar olingo lacks a prehensile tail and is more carnivorous.

Civets

Excluding the mostly diurnal mongooses and the predatory nocturnal species included above, the other members of the family *Viverridae* (about twenty-two species) are omnivorous. They include the nocturnal palm civets and the civets, which have been providing musk since prehistoric times. The civets' anal scent glands are highly developed, and secrete a fatty yellow substance with a musky odor which they use for marking territorial boundaries, for communication, and for defense. This secretion also has commercial value and is still collected from animals of the genera *Viverricula*, *Civettictis*, and *Viverra* for the perfume industry, where it is known as "musk" or "civet." To expand this industry civets have even been introduced and established in various countries including the Philippines and the Comoro Islands as providers of musk. Civets are also eaten, especially in China, where infected animals may have been responsible for the 2003 SARS outbreak. The most unusual use of civets, however, is the production of kopi luwak, an Indonesian coffee made from raw ripe beans that have been eaten by caged common palm civets and passed through their digestive tracts, when only the outer fleshy covering is removed. The nonhunting civets and palm civets have good senses of hearing and sight, and of course a good sense of smell, which is to be expected of animals that have evolved scent glands for their own use.

Small Indian Civet (*Viverricula indica*)

This is a mainly terrestrial civet which is also believed to excavate its own burrow for nesting, and which prefers to escape threats by running rather than climbing, although it climbs with ease. It is a long and slender animal, with an overall length of 39 inches (1 m) and a maximum weight of 7½ pounds (3.5 kg), which lives in a wide range of habitat from grasslands to forests in southern Asia and Indonesia. It has a base color of buffy-brown, and is covered with longitudinal lines of small spots on the forequarters and larger, more elongated markings on the hindquarters. Its legs and feet are black and its long tail is ringed with black and white. Both males and females have musk glands and are kept caged for its regular extraction, although apparently more for local consumption than for international supply. It has also been introduced into the Philippines, the Comoro Islands, and Madagascar for musk production. This civet is an omnivore, a hunter of small mammals, lizards, and birds, which also eats carrion and fruit.

African Civet (*Civettictis civetta*)

The largest member of the family and a totally terrestrial animal, the African civet lives in the grasslands and woodlands of Africa south of the Sahara. It is the most dog-like of the viverrids, with a heavy body, slender legs, and a bushy tail. Its thick coat of coarse black fur is covered with yellowish-white spots and bars, and its head is pale with black cheeks. This large civet weighs up to 44 pounds (20 kg), and its perineal gland secretions are collected in Ethiopia for the perfume industry, which is still unfortunately based upon the capture and caging of wild animals rather than any attempts to farm them. It is an omnivore, an active predator of rodents, birds, and invertebrates that also preys on small domesticated livestock such as chickens and goat kids, and eats carrion and plant matter, especially fruit. It is believed to dig its own burrows for birthing and pup raising.

Binturong (*Arctictis binturong*)

The largest arboreal viverrid, and the only prehensile-tailed one, the binturong weighs up to 30 pounds (14 kg) and has a tail 30 inches (76 cm) long, almost the length of its head and body. It has a long and coarse coat of shiny black hair tinged with grayish-buff, its head is speckled with gray, and it has white whiskers and ears fringed with white. The binturong lives in the equatorial rain forests of mainland Southeast Asia, Indonesia, and the Philippines, where it climbs well but ponderously. It also swims and is apparently able to catch fish, but it is a general omnivore that eats fruit, flowers, and nectar, and usually relies upon more easily obtained items like the eggs and young of tree-nesting birds for its animal protein, as it is hardly quick enough to hunt active animals. It is mainly a solitary animal, but both parents may remain together during the cub-raising period.

Notes

1. Although a carnivore is any flesh-eating animal, to avoid confusion the name is generally reserved for members of the mammalian order *Carnivora*, while other meat-eaters—such as Tasmanian devils, false vampire bats, eagles, and crocodiles, are preferably called carnivorous animals, rather than carnivores.

2. From a single source in the Paleocene Epoch between 55 and 65 million years ago.

3. Although they are hunters, the dogs also eat plant matter.

4. A small U-shaped bone in the throat supporting the vocal chords.

5. The upper premolar and lower molar teeth of the carnivores, adapted for cutting flesh.

6. The free-ranging "wild" bison in Wood Buffalo National Park and Yellowstone National Park.

7. After the *Viverridae*, which has seventy-one species.

8. Unlike the diurnal grizzly bear, which can put on a remarkable turn of speed to run down an elk calf.

9. The other major group of procyonids are the coatimundies, which are all diurnal.

10 Fair Game

The large nocturnal predators rely for their survival on the hoofed mammals or ungulates, the grazers and browsers that convert masses of herbage into meat and are the most abundant source of protein on land. Yet most ungulates are diurnal, and there are insufficient nocturnal species to meet the predators' needs. Consequently, when resting and digesting, and in most species ruminating, at night after cropping grass and plucking leaves all day, these animals—zebra, gnu, giraffe, eland, and many other antelope—are the prime targets of Africa's large carnivores—lions, leopards, and spotted hyenas. At night in Asia, nilgai, axis deer, and bar-asingha must be wary of tigers and leopards. The diurnal prey species are at a distinct disadvantage at night as the nocturnal predators' senses are more attuned than most prey species for nighttime activity. The following day, when the ungulates are refilling their stomachs with herbage they must again be watchful for predators, which may also hunt in daylight, especially if they were unsuccessful the previous night. The smaller nighttime predators like the caracal and serval in Africa seek appropriately sized prey animals, some of which are mainly also active at night, such as the dik diks and duikers, plus small diurnal antelope like the klipspringer when it is resting. In Southeast Asia clouded leopards and golden cats hunt the tiny mouse deer after dark.

Excluding the rodents, which are the main prey of the smaller predators,[1] the second-largest group of meat providers for the wild cats and dogs are rabbits and hares, which are more plentiful in many regions than ruminants, and although of considerably smaller body weight, provide an adequate meal for a cougar or dingo. Surprisingly, primates also provide several predators with part of their protein intake. Baboons and monkeys—including some arboreal species—are a favored prey of the wild cats, even large ones.

■ UNGULATES

The ungulates once all belonged to the old order *Ungulata*, hence their well-known and still-used common name, but they are now divided according to their hoof structure, specifically whether they have an odd or even number of toes. Those with an even number of toes are members of the order *Artiodactyla*, while the odd-toed ungulates are classified *Perissodactyla*. Consequently, they are often respectively called artiodactyls and perissodactyls, or collectively just herbivores, but the original "ungulates" is still acceptable.

The perissodactyls are the diurnal horses, zebras, and asses in the family *Equidae*, the diurnal rhinoceroses in *Rhinocerotidae*, and the nocturnal tapirs in *Tapiridae*. Adult rhinos are safe from the large carnivores, but their babies are vulnerable to lions and tigers, and the tapirs and zebras are major targets for several nocturnal predators.

The artiodactyls are the most plentiful mammalian herbivores, grouped in three major lineages, the *Suiformes*, *Tylopoda*, and *Pecora*. The *Suiformes* contains the hippopotamuses, wild pigs, and the pig-like peccaries, all of them nonruminants. The hippos are nocturnal and only their young are vulnerable to lions and spotted hyenas. The pigs and peccaries are also mainly nocturnal and are major food items of all the large cats in Africa, southern Asia, and South and Central America. Within the *Tylopoda* are the ruminating camels and llamas, and the latter's wild ancestors the vicuna and guanaco, all diurnal animals, of which only the guanaco is a major prey species—of the cougar in southern South America. The *Pecora* contains most of the ruminants, including the cattle, antelopes, deer, giraffes, sheep, and goats—which are the major providers of protein for the wild cats, dogs, and hyenas.

It is obvious from studying the head of an antelope or deer that the large ears, nostrils, and eyes are adapted for the receipt of information, and that it is a defensive, not an offensive animal. The herbivores are almost equally dependent upon the senses of sight, hearing, and smell, and their sensory stimuli—which are relayed to the brain's neocortex as electrochemical impluses—trigger the rapid muscular action needed to propel them into flight.

Unlike the predators with their foward-pointing eyes and overlapping binocular vision, the prey animals' eyes are usually situated on the sides of their head giving them a wider field of vision—frontally, at the sides and even to the rear. Thus their sight is monocular, each eye having a separate field, and their pupils are horizontal and widely oval in shape for scanning a wide area in the horizontal plane. However, their range of visibility at night is necessarily short and sight is therefore less effective than during the day, so more reliance is then placed on hearing and smell.

Flight is the ungulates' first defense, and they are built expressly for it. They walk on their toes, which are shod with small hard hooves, some small defense against predators, and their legs have evolved to carry the body from danger at high speed. Their running powers depend upon the type of surface they have evolved to occupy, their number of toes, and how their body weight is distributed. Those with split hooves (the antelope and deer), where the weight of the body passes between the third and fourth digits,[2] can run faster on uneven surfaces, such as rocky desert, soft soil, and soft sand, but they are not as fast on hard surfaces. In contrast, the single-toed animals (the horses and asses), in which the weight is borne on the third toe,[3] which is enlarged into a hard hoof, can run fast on flat hard surfaces, but find it heavy going on softer ones.

The flight distance is the critical point at which an animal will flee in response to the approach of a perceived threat. The reduction of their flight reflexes is the essense of taming an animal to accept the closeness of humans. The flight reflexes of prey animals are highly developed and work instinctively to place a safe distance between them and their predators; speed is the most important means of escape when a predator enters the ungulates' field of flight. The distance which they allow between themselves and the predator is based upon the species involved and their past experience of attack. They also know when predators are in a hunting mode, and grazing antelope may allow lions to pass by quite closely during the day when they are not hunting.

Flight is powered by the rush of adrenalin which immediately speeds up the heartbeat, circulatory rate, respiration, and muscular action. This furious rate of metabolism is evidenced by the flaring nostrils, staring eyes, and trembling body and legs. When flight is impossible for some reason and the intruder enters the animal's shorter fight distance, then it may defend itself, but most ungulates are not equipped to do battle with a predator. Most do have horns or antlers, although only in the males in many species, and although they appear to be excellent defensive weapons, their primary purpose is to resolve disputes of territory and sexual dominance, and few animals use them against predators. Those that do stand up to predators include the sable, roan antelope, and oryx, which have fought off and even killed attacking leopards and individual lions. Musk oxen make a determined effort to protect their calves from wolves, and the mountain goat has used its short but sharp horns effectively against cougars and wolves. It is interesting to note that these are all species that have horns and not antlers, and in which both sexes are horned. If they were essential for general defense, then it follows that the females would also have them. All female deer, except the reindeer and caribou, are antler-less, yet it is often repeated that deer use their antlers to defend themselves against predators. Situated as they are frontally, protecting the head and neck, antlers may offer some protection against a single predator, and by sweeping them side to side may deter a solitary carnivore, but they are of little defense when in flight with a pack of wolves or wild dogs snapping at their bellies. If they provided such good protection from predators the deer of the world would soon be restricted to a population of antlered males.

Hiding from predators is not an option for most ungulates, certainly not for the wildebeeste, eland, and zebras resting at night on the open African savannahs. Species that rest in the bush are perhaps a little safer, and the complicated stomach of the ruminants for digesting plant fiber is actually an adaptation for predator avoidance, as it allows them to quickly ingest a large amount of food in an exposed position, which can then be ruminated under cover in comparative safety. But a white-tailed deer resting in the forest can be located by a cougar, and in Africa

Gemsbok *A purely diurnal desert antelope, the gemsbok of southwestern Africa is preyed upon by lions and leopards when resting at night, but it is one of the few prey species that defends itself vigorously, using its rapier-like horns very effectively, even against the large cats.*
Photo: Courtesy Harcourt Index

duikers may dash through the forest quicker than any predator when alarmed, but they are still ambushed by leopards and African golden cats. Mouse deer, the duiker's forest counterpart in Asia, similarly run for their lives when alarmed, but they are not safe from expert ambushers like the clouded leopard and the Asiatic golden cat. For the prey animals that are nocturnal, such as the hippopotamus whose calves are vulnerable to lions and hyenas when they are grazing at night, resting in the water during the day is not too risky as no large land predator would have access to the calves then and the herd provides protection from crocodiles.

Many of the prey animals are gregarious as there are certain advantages to living in a group. Line breeding from the strongest male ensures improvement of the senses and survival of the fittest, and more eyes on the alert means better warning of approaching predators; in the case of zebras and musk oxen, greater ability for the herd to defend its young. Herd life does not reduce the overall risk of predation, however, as hunters will kill that night or within the next few days whether their victim is solitary or selected from a group; but there is the advantage to the herd that the carnivores can pick the weakest animal when a choice of victim is possible, thus assisting in maintaining the vigor of the herd.

Some of the Species

Odd-toed Ungulates (Perissodactyla)

Burchell's or Plains Zebra (*Equus burchelli*)

Zebras are wild horses, the striped African relatives of those on the Russian and Mongolian steppes that gave rise to all the breeds of domesticated horses. They have vertical stripes from the head to mid-body where they begin to turn toward the tail and become horizontal on the rump. They are animals of the grassy plains and open woodlands and bush; they have the most specialized mammalian foot development for rapid flight in open country, and are strong runners with great endurance. They

Burchell's Zebras   *Africa's diurnal striped horses are fair game for lions, leopards, and hyenas at night, when their senses are no match for the predator's highly attuned night vision and senses of hearing and smell.*
Photo: Clive Roots

have lateral vision and can therefore see approaching predators at the left and right outer boundaries of the visual field. Burchell's zebras are compact horses, standing 55 inches (1.4 m) at the shoulder and weighing up to 825 pounds (375 kg). Although broad and well-spaced rump stripes are characteristic of the species, there is considerable variation in the several races, plus geographic differences and individual aberrations in the striping. Unlike the more northerly, narrow-striped Grevy's zebra *Equus grevyi*, their stripes go around the belly, meeting at the mid-ventral line, and occasionally reverse patterning is seen, in which the individual has a black coat with white stripes. The zebras' major predators are lions and spotted hyenas, which hunt them at night and occasionally during the day. As zebras cooperate in defending the family and protecting the foals, the normally solitary spotted hyena has learned to band together in packs to hunt them.

Brazilian or South American Tapir (*Tapir terrestris*)

The commonest of the three[4] species of neotropical tapirs, the Brazilian tapir also has a much larger range, the northern half of South America east of the Andes—from the Caribbean coast to southern Brazil and Paraguay. It is a large, compact animal, with short and sturdy legs, and stands 48 inches (1.2 m) at the shoulder and weighs up to 660 pounds (300 kg). Its body is barrel-shaped with a tapering front and long, narrow skull that is convex in profile, a long and thick neck, short tail, and small eyes. Its snout is elongated into a short but flexible proboscis—an adaptation for browsing. Tapirs have thick skins, a coat of short and sparse dark-brown hair, and a characteristic neck mound topped with a short mane. Their forefeet have three main digits and another small one which only reaches the ground when the going is soft.

The tapir is semiaquatic and is rarely found far from water, preferring river-bank bush where it has an escape route from jaguars, as it is a better swimmer than the large cats. It shelters by day in thick undergrowth and comes out to feed at dusk. As the New World's largest land mammal within its range, it is the favored prey of the larger predators—the cougar, jaguar, and humans. A browser, its diet comprises leaves, shoots, fruit, and seeds, and it is considered one of the rain forest's most important seed dispersers, distributing undigested seeds over its wide territory.

Even-toed Ungulates (Nonruminants) (Artiodactyla)

Wild Boar (Sus scrofa)

Known until recently as the European wild boar, this species was indeed misnamed, as it has a much larger distribution than simply Europe, ranging across the Palaearctic Region from south of the tundra zone to the Middle East, India, the Malay Peninsula, and Sumatra, plus North Africa. It, or its domesticated descendant, has been introduced and is established in several countries including North America, Australia, and New Zealand. Domesticated since 5,000 B.C., selective breeding has produced a wide range of breeds, the largest weighing over 880 pounds (400 kg). The typical wild boar is a large, dark grayish-brown to black animal with a sparsely haired, stiff and bristly coat, standing just over 39 inches (1 m) at the shoulder, with a body length of 60 inches (1.5 m) and weighing up to 550 pounds (250 kg). It has short legs, a stocky body, a long, truncated snout ending in a cartilaginous disc enclosing the nostrils; and a long, sloping skull, small eyes, and long ears. Although wild boars have four toes on each foot (the first toe is missing) they walk only on the middle two (the third and fourth) which are hoof-like, and the remaining two are tiny and do not reach the ground, although they are also hoofed.

The wild boar is mainly nocturnal, and has poor sight, but acute senses of smell and hearing. It lives in a wide range of habitat including woodland, grass-lands, and reed beds, but is seldom found far from moist areas as it loves to wallow. It is omnivorous, and eats virtually anything of plant and animal origin; the domestic breeds are well known for their ability to eat human garbage, and humans. Wild boars have numerous predators in their natural range. Tigers, leopards, striped hyenas, wolves, and dholes are their major hunters, but piglets are prey to many smaller carnivores including the Eurasian lynx and jungle cat.

Wart Hogs *Normally a diurnal pig, the wart hog is now nocturnal in many regions due to excessive persecution, but whenever it is active it is a favored prey species of the lion, leopard, and spotted hyena. The piglets are especially vulnerable when a hyena draws the mother's attention, allowing others to sneak in and seize her babies.*
Photo: Steffen Foerster, Dreamstime.com

Wart Hog (*Phacochoerus aethiopicus*)

The wart hog is a very long pig, with a head and body length of 60 inches (1.5 m), a shoulder height of 30 inches (75 cm), and weighs about 220 pounds (100 kg). Males have prominent warts on the sides of their heads and in front of the eyes, which are skin growths unsupported by bone, and which are less pronounced in the females. The wart hog has a very strong snout or rhinarium and can dig into the hardest soil; it is believed to be more herbivorous than the other pigs,[5] but it also eats locusts and carrion, and often feeds while leaning forward on its wrists. Wart hogs have the most impressive tusks of all the pigs. Their upper canine teeth turn upward as tusks, either after entering the mouth or by breaking through the skin of the snout, and normally then curl back toward the head. The lower canines also usually grow out beyond the lower jaw, rubbing against the upper tusks and creating a cutting edge on both. The wart hog lives in sub-Saharan Africa where it prefers grasslands, bushveldt, and wooded savannahs. Although it is a social species it generally occurs in separate groups of adult females or young males. It was the only truly diurnal wild pig, but nocturnal behavior has been forced upon it in many regions by excessive persecution. Abandoned porcupine or aardvark burrows are the warthog's favored sleeping places, but when outside foraging it is at the mercy of a number of predators. Lions, leopards, spotted hyenas, and African hunting dogs all prey upon them, and their piglets are vulnerable to ratels, caracal, and jackals.

Collared Peccary (*Tayassu tajacu*)

Peccaries are pig-like animals that take the place of the swine in the New World, and they are mistaken for pigs because they do resemble them. They have a similar shape, the same long and mobile head ending in a cartilaginous, disc-like snout, and a coat of coarse hair, but their bodies are much narrower and their legs are longer and slimmer. A major difference, however, is the presence of a scent gland just above the rump that emits a musky secretion when the animal is excited, and is used during their social behavior and for territorial marking. They rub their throats against the glands of their companions, creating a communal scent. They also differ from the pigs in the structure of their feet, the third and fourth digits of the hind feet being united at their base like those of the ruminants; and in their teeth, in which the upper canines grow downward as short tusks, not upward like the pigs. The collared peccary is the smallest of the three species,[6] and the sexes are similar in size, about 39 inches (1 m) long, 20 inches (50 cm) high at the shoulder, and they weigh about 66 pounds (30 kg); the coat is dark-gray with a narrow white collar.

Peccaries are gregarious animals that live in herds, or sounders, of up to fifty animals, which maintain contact at night through grunting, as they have poor eyesight, but their hearing and smell are acute. Their habitat ranges from the deserts of the southwestern United States to the tropical rain forest of Amazonia, where they are a favored prey of the jaguar, cougar, and man, although they are dangerous adversaries that fight courageously with their tusks. Their young are watched carefully by the adults, but are vulnerable to ocelots. Peccaries love to wallow and their muddy holes in the forest floor create habitat for amphibians. They are quite sedentary animals and their natural diet is fallen fruit and nuts, shoots, roots, invertebrates, and occasionally carrion.

Hippopotamus (*Hippopotamus amphibius*)

The hippo is the largest semiaquatic land mammal, with males reaching a length of 16 feet 6 inches (5 m), standing 63 inches (1.6 m) high at the shoulder, and weighing 5,500 pounds (2,500 kg); females are a little smaller. They have heavy bodies, short legs, a wide muzzle and slit-like nostrils, and are so sparsely haired that their brown skin, which is tinged coppery-purple, can be seen beneath. Glands in their skin secrete red-pigmented droplets, giving the appearance that they are sweating blood. Originally with a much larger range, the hippo has for many years been restricted to Africa south of the Sahara, although extending north along the Nile Valley to Khartoum. It is now scarce in West and South Africa, as a result of habitat loss or hunting for its hide, meat, and teeth. Hippos are nocturnal and sleep all day in water; they come ashore at dusk to crop grass and young reeds with their horny lips, walking along well-trodden pathways to their grazing grounds, which may be 1 mile (1.6 km) from the water. They have a compartmentalized stomach and "fore-gut or pre-gastric fermentation" of herbage similar to the ruminants, but they do not ruminate. Hippos are very aggressive animals, especially mothers with young, and are considered the most dangerous of Africa's large game species, responsible for many human deaths. All predators avoid healthy adult hippos, however, both lions and spotted hyenas take babies and old and sick individuals. When they are resting by day in the water they are safe from land predators, and crocodiles generally do not attack young protected by the adults.

Even-toed Ungulates (Ruminants) (Artiodactyla)

Common Eland (*Taurotragus oryx*)

The diurnal eland, the largest antelope, is an ox-like animal with a grayish-fawn coat with cream-colored vertical stripes on the upper body. It has a hump on the withers, a dewlap framed with long black hair, and both sexes have heavy spiral horns. Males stand 70 inches (1.8 m) high at the shoulder and weigh up to 2,000 pounds (900 kg), and females reach a height of 56 inches (1.4 m) and can weigh up to 1,000 pounds (454 kg). The eland is a very social animal that lives in herds on the plains and wooded savannah of eastern Africa from Ethiopia to South Africa, where it is basically a grazer of coarse herbage, but also browses the leaves of trees and shrubs. It is the only antelope to have been domesticated, mainly for its milk which has triple the fat content and double the protein of cows' milk, and is richer in calcium and phosphorus. With good nutrition the domesticated eland cows in the long-established herd at Askaniya Nova in the Ukraine lactated for up to 300 days, some animals providing almost 2 gallons (9 L) per day. Also, although most African antelope have subcutaneous fat, eland meat has marbling throughout the flesh like beef, and remains relatively tender into old age. In Africa several predators also consider the eland fair game and it is hunted at night by lions, leopards, spotted hyenas, and by African hunting dogs on moonlit nights (they are sight and not scent hunters).

Nilgai or Blue Bull (*Boselaphus tragocamelus*)

The nilgai is India's largest antelope, a diurnal animal of the plains and open woodland, ranging throughout the subcontinent south of the Himalayas, but excluding the extreme southern tip, and extending westward into Pakistan. Males are dark bluish-gray with white underparts and inner ears; they stand 55 inches (1.4 m) at the shoulders and weigh up to 600 pounds (270 kg). The shorter females, standing about 43 inches (1.1 m), are orange-brown in color and weigh up to 450 pounds (200 kg). Males resemble the females until they are almost a year old, when they begin to attain their steel-gray coloration, but this is not completed until they are at least three years old. Both sexes have a neck mane and the bulls have a throat tuft about 6 inches (15 cm) long, and short horns. During the breeding season the bull's coat darkens and his neck thickens, resulting in a very pronounced slope from head to rump. Nilgai browse and graze, depending upon food availability, and are able to reach very high for leaves when they raise up on their hind legs. The males are aggressive animals, but they are no match for Bengal tigers and leopards, especially when they are attacked in darkness, nor for a pack of hunting wolves or dholes.

Brindled Gnu or Wildebeeste (*Connochaetes taurinus*)

An unusual diurnal antelope, in which the back slopes down from the withers to the rump, the wildebeeste stands 57 inches (1.45 m) at the shoulder and weighs 600 pounds (275 kg). It is grayish-silver with a black face, beard, mane, and tail, and both sexes have upward-pointing horns that curve inward. Gnu have a wide distribution from southern Kenya to northern South Africa, where they graze the open, grassy plains and wooded savannahs and are the dominant large mammals on Tanzania's vast Serengeti Plains. They usually live in small herds of about thirty animals, with several herds controlled, and patrolled, by territorial males. The herds follow seasonal grass growth, moving on when areas are depleted, and predicting the rainfall; they are famous for their mass migrations, which become virtual death marches when their route takes them twice annually across rivers where Nile crocodiles lie in ambush. Gnou (their Hottentot name) are seasonal breeders, with all births occurring about three weeks before the rains begin, so the calves have access to fresh grass when they are weaned. Birthing time is a period of intense excitement for predators, which circle the herds seeking an opportunity to seize a calf before it can run; but they must be quick, because calves stand within a

Impalas *The common antelope of the grasslands and bushveldt of South and East Africa, the diurnal impala is a major prey species of lions, leopards, and spotted hyenas by day and night, and its calves are taken by ratels, jackals, and caracals.*
Photo: Clive Roots

few minutes of birth and are soon able to follow their mothers. Their major nocturnal predators are the lion, leopard, and spotted hyena, and newborn calves are also vulnerable to jackals and ratels.

Impala (*Aepyceros melampus*)

This is one of the most common antelopes of the bushland and acacia savannahs from Kenya south to northern South Africa and westward into Angola. It is a very recognizable species, with dark reddish-fawn upperparts and contrasting paler sides, and then a white underbelly, each color very distinctly separated. In the females the upperparts are a pale brownish-fawn, and both sexes also have a black line on their hindquarters. Adult bucks stand 39 inches (1 m) to the shoulder, weigh 143 pounds (65 kg), and have lyrate horns up to 36 inches (92 cm) long, which are ridged on the outside; the smaller females are hornless. They are very agile and swift animals; they run fast and leap high, and are able to jump 10 feet (3 m) vertically and 33 feet (10 m) horizontally. Unlike all other antelope they have a large, brush-like tuft of coarse black hair covering the scent gland above the heel on each hind foot. Impala are diurnal antelope; the males are solitary and territorial and the herds of females, up to thirty strong, cross in and out of males' territories. The males vocalize with a loud bark when protecting their territory; they fight by rushing at each other and clashing horns, and such battles may be fatal despite their very thick hides. Births occur to coincide with the annual rains, and the calves are taken by jackals, caracal, and ratel, while the adult impalas are preyed upon by lions, leopards, and spotted hyenas at night and African hunting dogs by day.

Yellow-backed Duiker (*Cephalophus sylvicultor*)

The duikers are small to medium-sized, mainly nocturnal antelope, of which there are nineteen recognized species that are endemic to sub-Saharan Africa. They have low-slung bodies, short legs, arched backs, and a wedge-shaped head. Their habitat includes bush country, dense moist forest, and swamps, and when threatened they emit a shrill whistle and quickly dive into the undergrowth. Duiker are predominantly browsers, but they eat fallen fruit and even scramble into low bushes to pick their own. They also eat insects, small vertebrates, and carrion, and are therefore the most omnivorous bovid–hollow-horned ruminants of the "cow" family *Bovidae*. The yellow-backed duiker occurs in the whole block of equatorial forest from Gambia to Kenya, and is mainly a solitary animal; the adult males are very territorial and do not tolerate others in their space. It is the largest duiker, reaching a weight of 165 pounds (75 kg) and is a spectacular animal with a dark coat and bright-yellow rump patch, which it can erect when alarmed. Its horns are short and pointed and are curved backward. A nocturnal animal, it hides by day on a regularly used bed of vegetation in dense undergrowth or among the roots of a large forest tree. At night it must evade leopards and African golden cats, and its calves are vulnerable to the large African civet.

Bushbuck (*Tragelaphus scriptus*)

The bushbuck is a nocturnal antelope of the forest edges, montane forest, riverbanks, and gallery forest, throughout sub-Saharan Africa, rarely far from water. It is a medium-sized antelope with a shoulder height of 34 inches (86 cm), and weighs about 130 pounds (59 kg). Its coat color is variable, from chestnut to dark brown, with white patches or spots in irregular rows on the sides and rump, and a white band at the base of the neck. Females are smaller and lighter, and only the males have horns, which are short and slightly spiralled, and about 18 inches (46 cm) long. Bushbucks are solitary creatures, most active at night and early in the morning in regions where they are regularly disturbed. Their reaction to threats varies, from standing perfectly still, dropping down to lie flat or bounding away barking as they go, but with their hunched gait they are slow and clumsy runners, and rarely outpace leopards. Bushbucks are mainly browsers, of leaves, shoots and flowers, acacia pods, and tender bark, and they often associate with baboons, picking up the fruit that they drop, despite the baboon's habit of seizing newborn bushbuck calves. Their major predators are lions, leopards, and spotted hyenas, and they are also vulnerable to serval, golden cats, and ratels.

Musk Ox (*Ovibos moschatus*)

Musk oxen have characteristics of both cattle and sheep but are believed to be closer relatives of the takin, a large, golden-brown herbivore of the eastern Himalayas and southern China. They are probably the toughest, roughest, and hardiest of the northern land mammals, possibly of all land mammals. Native to the tundra of Alaska, Canada, and Greenland, they are protected by their coarse outercoat of darkish-brown guard hairs and insulating inner lining of dense and soft light-brown wool called qiviut, and are impervious to the coldest weather. Massive in form, their bulk is enhanced by a shoulder hump and the long, thick coat that reaches to the ground, and both sexes have pointed curved horns. Adult bulls can weigh up to 1,325 pounds (600 kg) and stand about 52 inches (1.3 m) at the shoulder. Their ponderous appearance is misleading, however, for they move swiftly and with agility when necessary, and form circles around their calves to protect them from wolf attack.

Musk oxen seek moist habitat in summer, usually in river valleys or along lake shores where they graze on grasses and sedges, and swim well when necessary. For the winter they migrate to more exposed areas where the wind blows the snow off their foods—mainly ground-hugging browse such as willow and rhododendron. Freezing rain is hazardous for them; it accumulates in their coats and chills them, affecting their mobility and effective defense against wolves, their major predators.

Barbary Sheep or Aoudad (*Ammotragus lervia*)

A goat-like sheep, the aoudad has glands under its tail like the goats, but lacks their chin beards and smell. It has a short and smooth, sandy-colored coat, with its underbelly and insides of its legs whitish. The male has a well-developed throat mane and hairy "chaps" on the forelegs. The aoudad has hybridized only rarely with domesticated goats and sheep. It occurs naturally in the mountain ranges of North Africa from the Atlas Mountains to the Red Sea, and there are at least six recognized races within that range. Adult aoudad rams stand 38 inches (96 cm) at the shoulder and weigh up to 300 pounds (136 kg), while females are almost as high but only half the weight. The male's horns form a sideways curl, are more open than those of the true sheep, and can reach 30 inches (76 cm) along the curve. Aoudad are very adaptable and opportunistic animals, surviving on scant vegetation in the wild, and needing little water. In fact, they can survive if necessary on the juices of the plants they eat, plus dew, for weeks at a time. Their hardiness, appearance, resistance to parasites, and tolerance of overcrowding has made them popular zoo animals, and they have been controlled by man for so long they are now considered domesticated. However, the unknown subspecific origin of the founders of captive herds and their long-term, uncontrolled breeding has resulted in a population of generic domesticated aoudads, plus flocks of feral animals in Texas where they have been released or escaped and are now well established. In their natural habitat the adult aoudad's only major predator is the leopard, but lambs would be vulnerable to caracal and eagles.

Roe Deer (*Capreolus capreolus*)

Roe deer are small, attractive deer, with a reddish coat in summer, which becomes grayish-fawn in winter, and a white throat. They have no visible tail, but in winter the females grow an anal tush—a tuft of hair that resembles a tail. There are three recognized races. The European roe deer (*C. c. capreolus*), of England, Scotland, Europe, and Asia Minor, is the smallest, standing 29 inches (75 cm) at the shoulder and weighing 48 pounds (22 kg). The Siberian roe deer (*C. c. pygargus*), which inhabits Russia from the Urals to the Pacific, in the southern half of the forest zone, is taller and sturdier, standing 35 inches (88 cm) at the shoulder and weighing up to 88 pounds (40 kg). The third subspecies, from China, is slightly smaller. Roe deer are mainly animals of the forest and secondary growth, and are predominantly browsers, although they also venture out to graze on young grasses in forest clearings and on open grassland adjoining the forests. Bucks shed their antlers in October or November, and their new growth is hard and clean of velvet by the end of April. From then they have a well-deserved reputation for aggressiveness to the does during the rut, chasing them continuously and often injuring them with their spiky antlers. They have many predators, nocturnal ones being the tiger, leopard, wolf, dhole, and Eurasian lynx; and red foxes and badgers can easily kill newborn fawns.

Roe Deer *Forest-dwellers of the northern Palaearctic Region that come out at dusk to graze in clearings, roe deer are vulnerable to many predators. They are hunted by wolves and the European lynx, while in the Far East their major predators are the Siberian tiger and the North China leopard.*

Photo: Adrian Jones, Dreamstime.com

Elk or Wapiti (*Cervus elaphus*)

The North American races[7] of the red deer (*Cervus elaphus*) are known as elk or wapiti, and they are the second-largest deer on the continent after the moose. Adult bulls stand 5 feet (1.5 m) at the shoulder and weigh 990 pounds (450 kg). In summer they have short, bay-colored coats with darker legs and head, and a large, pale caudal patch, but in winter their coats are much coarser and darker and the bulls have a dark mane. Elk are both nocturnal and diurnal, and were once very common throughout North America, but were heavily hunted in the late nineteenth century, initially for their hides and then solely for their canine teeth or tusks, which were in demand by members of Elk's Lodges. They were reduced to just 10 percent of their former numbers, but they have recovered in recent years with protection and their relocation to many refuges. Elk eat a wide variety of vegetation, depending upon their habitat and the available herbage. However, it appears that given the opportunity they prefer grasses, but are forced to browse when deep snow prevents them from reaching grass. They love to wallow in mud, especially the bulls, when they are rutting. Their only major nocturnal predators[8] are cougars and wolves, but lynx and wolverine can kill newborn fawns. Elk were introduced into New Zealand's Fiordland National Park in 1905, and thrived in the cool, wet climate to the west and northwest of Lake Te Anau. But this region was also a stronghold of the previously introduced red deer and hybridization has since resulted in the elk's deterioration.

Red Deer (*Cervus elaphus*)

Red deer are widely distributed across Eurasia and are the only true deer occurring naturally in Africa, where they occupy Tunisia and northeastern Algeria. They have been successfully introduced into Argentina, New Zealand, and Australia. Red deer occur in a variety of habitat, including grasslands, temperate forests, alpine forests, and even very arid regions. Their wide natural distribution has resulted in the evolution of almost twenty distinct subspecies, which differ considerably in body and antler size. Even within Europe they vary from the smallest race, which lives on the islands of Corsica and Sardinia where the stags weigh only about 250 pounds (113 kg) and have small antlers, to the largest red deer in Poland and Romania, where stags have large antlers and may weigh 640 pounds (290 kg), rivalling the elk in size. Races of the red deer also occur in central and eastern Asia, in an arc from the Tien Shan Mountains to the Pacific coast of southern Russia, with two isolated forms in Tibet and central China. They have very unspecialized feeding habits, grazing grasses and a wide range of forbs, lichens, mosses, and fungi, and browsing on trees and shrubs. Like the elk, red deer are both nocturnal and diurnal in their activities.

Both sexes of red deer bark when alarmed, but during the rut the stag roars like a lion, and the hinds grunt. After rutting, the old males go off on their own while the hinds stay in herds which are usually quite permanent and also contain young males. The number and variety of their predators increases eastward. In Great Britain the only natural ones are the diurnal golden eagle and the red fox, which are both capable of taking newborn fawns. In the more remote parts of Europe wolves are still a major predator of mature deer and farther east still the nocturnal big cats—snow leopards, leopards, and tigers—are all potential predators; they are also sought by wolves and dholes. Fawns, which are left by their mothers in a secluded place for several hours between feeds, lack any scent that could attract predators, so they are difficult to find, but later they are vulnerable to lynx and the Asiatic golden cat.

White-tailed Deer (*Odocoileus virginianus*)

The most common New World deer, the white-tailed deer has been hunted extensively by the native peoples of North America and is now a prime target for sport hunters. It is widespread in a range of habitat from southern Canada through the United States and Central America to northern Brazil, excluding the west coast of Canada and the arid lands of the American Southwest. Within its large natural range the many races show considerable variation in size, from the large northern deer, which may weigh 330 pounds (150 kg) and stand 39 inches (1 m) at the shoulder, to the small Florida white-tailed deer of the Everglades and the Keys, which is just 26 inches (65 cm) high and weighs only 80 pounds (36 kg). The white-tail is reddish-brown in summer, brownish-gray in winter, and has a coat of brittle, tubular hairs. It resembles the mule deer, the most obvious differences being its tail, which is brown above and white below, and its ears, which are smaller than the mule deer's and are just half the length of its head. The white-tail's antlers have a single main beam and several smaller branches. In North America cougars and wolves are their major predators, and their fawns are taken by lynx, bobcat, coyote, and wolverine. In Central and South America they are also sought by cougars, plus jaguars and ocelots, and newborn fawns are vulnerable to the bush dog, tayra, and jaguarondi.

Barasingha or Swamp Deer (*Cervus duvauceli*)

There are two races or subspecies of this deer, *C. d. branderi* of central India and *C. d. duvauceli* of northern India and adjacent Nepal and Assam. The name swamp deer really only applies to the animals of northern India, which are seldom seen far from the river flats or marshy habitat, whereas the central Indian race has no access to swamps, and grazes on the open grassy plains. Barasingha have a golden-brown coat in summer that becomes dark brown in winter with a shaggy ruff, and males reach a height of 52 inches (1.3 m) at the shoulder and weigh up to 450 pounds (205 kg). They are primarily grazers, eating the lush grass of the wetlands or the drier grasses of the plains, although the coarsest species are only eaten when green. Even when browse is available they are so inflexible in their dietary preferences that they suffer when fires destroy the grass in the dry season, and when there is intense competition from other grazers, which include the axis deer and nilgai. They are mostly crepuscular in their habits, feeding at dusk and dawn and seldom staying out in view later than mid-morning. Indian Mynahs act as tick birds and as alarms while they are resting during the day. They are a favorite prey of the tiger, leopard, wolf, and dhole, and their fawns are vulnerable to striped hyenas and golden jackals.

■ RABBITS AND HARES

In addition to the large ungulates, which provide the bulk of the larger predators' protein requirements, there is another group of mammals that are also major providers of food for the smaller cats, dogs, and other carnivores. They are the rabbits and hares, which are often mistakenly thought to be rodents but belong to the order *Lagomorpha*. They are close relatives of the rodents, and are separated from them primarily on account of their teeth. Rodents have just two incisors in each jaw, whereas the rabbits and hares have four in their upper jaw and two in the lower jaw. The extra ones are hidden behind the main upper pair, and are virtually nonfunctional as they lack the cutting edge of the main incisors, which are open-rooted and grow continuously. Like the rodents they lack canine teeth.

The rabbits and hares are all terrestrial, and are mainly nocturnal and crepuscular in their behavior. They are very prolific animals, able to breed several times yearly, producing large litters after a short gestation period, and the young mature quickly. In the absence of controls they can multiply rapidly and become major pests, as they did when introduced into Australia years ago, causing great environmental damage and financial loss to the agricultural industry. Fortunately, in their natural habitat at least, there are many controls on their numbers, as they are a major source of food for many predators, diurnal and nocturnal alike, snakes, birds, and mammals. In the far north their populations fluctuate, peaking about every decade and then declining, a cycle which is believed to be initiated by a period of abundant food followed by an increase in predators and the inevitable crash of both as the prey declines, but other factors must be involved or it would not occur so predictably. A gradual population buildup over several years may therefore be involved initially; and the subsequent stresses of overcrowding may also play a part in their eventual decline.

Some of the Species

European or Old World Rabbit (*Oryctolagus cuniculus*)

This common and very widespread rabbit has a coat that contains both dark and pale hairs mingled to produce an overall brownish color with a faint black tinge. It has a buff patch on the nape and its narrow ears are trimmed with black at their tips; and when adult it weighs about 4 pounds (1.8 kg). A very adaptable animal, proven by its ability to rapidly colonize the countries to which it has been introduced, it is believed to have Spanish origins, and from there to have spread naturally over Europe and western Asia. The British Isles may have been the first of its assisted introductions, possibly taken across the English Channel in the twelfth century as a source of food. More recently it has been released in Australia, New Zealand, Chile—from where it has invaded neighboring Patagonia—and islands in the Atlantic and Pacific Oceans, where it has severely affected native plants and animals. The European rabbit prefers sandy soil, especially on hillsides, and appears at dusk to graze. It is a very social animal that lives in complex warrens, where the young are born, naked, and blind until they are ten days old. The rabbit's acceptance of control by man, at least when collected as youngsters and hand-raised, has resulted in its domestication and the development of all the current breeds of rabbits. It is a highly productive animal that provides food for many nocturnal predators, including the eagle owl, lynx, wild cat, feral cat, red fox, polecat, mink, marten, and stoat. In Australia it is now a major prey species of the dingo and native cat, and in Patagonia it is food for the cougar and for the colpeo, a large fox.

Eastern Cottontail (*Sylvilagus floridanus*)

One of the thirteen known species of cottontails—the common "rabbits" of the New World—the eastern cottontail occurs from southern Canada through the United States east of the Rockies, Central America, and northern South America. It is a small, brownish-gray rabbit, with a fluffy brown-and-white tail, white feet, and rusty nape patch; and does not change its coat to white for the winter like the snowshoe hare. It is mainly an animal of the forest, of open areas adjacent to forest, cultivated land close to bushy undergrowth, and along the edges of muskeg swamps. It also favors unmaintained cleared areas with tall weed growth. Whereas most rabbits burrow, the cottontails raise their young in a surface nest in a secluded place, although some females may dig a shallow scoop for their nest, which is made of grass and fur plucked from her stomach. They are nocturnal, and normally spend the day concealed in dense undergrowth, or beneath a pile of brushwood, but are occasionally seen during the day. Cottontails are very prolific animals, able to produce three or four litters annually, averaging four young after a gestation period of twenty-eight days. The young are blind at birth, but their eyes open at the end of their first week and they begin to leave the nest when two weeks old. They are important small-game animals for all the major land carnivores. Cougar, wolf, coyote, lynx, bobcat, skunk, badger, gray fox, fisher, mink, long-tailed weasel, and the short-tailed weasel or ermine all seek the cottontail, as do the larger owls such as the great horned owl, great gray owl, and the barred owl. In the tropical parts of its range the cottontail must also avoid the cougar, ocelot, jaguarondi, margay, and bush dog.

Cape Hare (*Lepus capensis*)

Despite its name, derived from South Africa's Cape Province, this hare lives throughout Africa, in its preferred habitat of seasonally dry and semiarid country—semidesert, bushveldt, steppe, and dry savannah. It also occupies the Iberian Peninsula, other parts of southern Europe, and across Asia to China. The cape hare is a pale brownish-gray animal flecked with black, with ruddy brown legs, a pale belly, and a black-and-white tail; adults weigh up to 5 pounds (2.2 kg). Typical of the hares, it has long, thin limbs and large ears, and relies on hearing and vision for warning and speed for escape. When it is angry the cape hare drums on the ground with its forefeet and stamps its hind feet. Also typical of the hares, it does not dig a burrow like the true rabbits, but hides in vegetation during the day in its nest or "form," using the same one each night, and emerges at dusk to feed. It is a grazer and browser of grasses, forbs, and shrubs. Young cape hares are born fully furred and their eyes open soon after birth. It is likely that they lack body scent while in the nest to prevent drawing the attention of predators that hunt by smell. In Africa the cape hare is a major source of food for the leopard, serval, caracal, black-footed cat, striped and spotted hyenas, ratel, and jackals. In Eurasia it is preyed upon by the eagle owl, lynx, golden jackal, and red fox.

■ PRIMATES

Although monkeys, and primates generally, are rarely considered prey animals, they are actually highly sought by several predators, especially cats that can climb into the trees after them or ambush them when they leave the trees briefly to drink or forage on the ground. Ground-dwelling primates are naturally at greater risk, even though they are all diurnal species; they may be seized during the day when they are feeding or taken at night when they are roosting.

Refuging or communal sleeping is common in primates; their refuge trees are chosen for their apparent safety from predators, but they do not provide complete security as several cats are very good climbers. In Asia the clouded leopard is a very agile, tree-climbing feline that regularly catches monkeys. In Sabah it is known to kill proboscis monkeys and in Thailand pig-tailed macaques and gibbons have fallen prey to its speed and agility. It lies in wait in the lower branches for monkeys to pass beneath or actively stalks them on the ground. In South America the jaguar, although not such a good climber as the clouded leopard, is also successful at hunting monkeys in the lower branches of thick-limbed rain forest trees. The ocelot, largest of the South American small cats, also adopts similar hunting techniques for spider monkeys. However, it is in Africa that the art of monkey-catching has been perfected by several carnivores, and a number of primate species are involved. The thorn scrub, veldt, and wooded savannahs—where wide expanses of grassland are studded with acacia trees and occasional rocky outcrops—are the home of several primates that are mainly terrestrial but seek the safety of trees or rock ledges at night, and when threatened during the day. Three primates that are food for several of Africa's many predators, especially leopards, are included below.

Some of the Species

Patas Monkey (*Erythrocebus patas*)

With its long, slender limbs and narrow body the patas monkey is the primate equivalent of a greyhound, built for speed. A terrestrial monkey of the acacia-studded plains of sub-Saharan Africa, it spends most of its time on the ground, except when roosting in trees at night; and if disturbed while feeding in a tree prefers to jump down and run, at speeds up to 35 mph (56 kph), outdistancing most pursuers. It has long and narrow feet but runs on its fingers and toes, not its palms. Its red coat, white whiskers, and white stockings and gloves give the patas monkey a military appearance, and it is also called the hussar or military monkey. It is an omnivore, its main diet being the leaves, flowers, and seed pods of the acacia, grass and roots, and locusts, birds' eggs, and nestlings.

The patas monkey has several strategies for nighttime survival. Unlike the baboons, which crowd into one large tree at night, adult patas monkeys generally roost one per tree, and they rarely use the same tree twice, although this is mainly because they are continually on the move, their ramblings across the plains in search of food taking them several miles daily. They generally live in small troops consisting of a single male and several females and their young, and before they settle down for the night the male surveys the surrounding plains for predators; if one approaches he calls and bounds away, drawing it away from the females and babies. Despite their survival strategies, however, living on the ground during the day in habitat so rich in predators is very dangerous, and they are caught by leopards, spotted hyenas, caracals, and jackals. When roosting at night in acacia trees the very agile and persistent leopard climbs up after them, and in their confusion and terror they may fall out of the tree, when the leopard's superior nocturnal senses win the day.

Chacma Baboon (*Papio ursinus*)

Baboons are Africa's large terrestrial primates with dog-like faces, huge-canine teeth, and bare and often colorful buttocks. The chacma baboon of South Africa, which may not be a full species but a race of the savannah baboon (*P. cyanocephalus*), is a dull olive-brown animal that lives in a variety of habitat including the fynbos of

Chacma Baboon   *A male chacma baboon stands guard over his troop of females and their young in South Africa's Kruger National Park. Even in daylight lions and leopards may ambush an unwary individual, and at night leopards climb the trees in which the baboons sleep and panic them into falling.*
Photo: Clive Roots

Cape Province, the mopane scrub of the veldt, and on the treeless higher slopes and crags of the Drakensberg Mountains. It is an omnivore, which supplements its diet of grass, leaves, roots, and fruit with locusts, rodents, and lizards, and also sneaks up on young antelope and other primates, seizing them or their babies and literally tearing them apart. Despite this behavior, however, baboons are very courageous animals, and when a troop is confronted by a predator during the day, even by a leopard or lion, the adult males actually move toward it and demonstrate aggressively, showing their leopard-size canine teeth, which usually discourages the carnivore. This calls for sneaky behavior by the predators, which lie in wait for the opportunity to seize a straggler. Even lions have waited in ambush to quickly grab a baboon and run before the troop can rally. Baboons living in the mountains find a cliff ledge for nighttime roosting, safe from predators, but in the savannahs and bushveldt, sleeping in large trees may be safer than sleeping on the ground, though it is not totally secure. Baboons in number may harass a leopard during the day, but they make no attempt at defense at night when roosting, when there is no rallying or collective protection, their only concern being personal safety. Leopards consequently have much greater success at night, and hunt them in the trees, and in response the frantic baboons move to the outer thinner branches which will not bear the predator's weight, nor their own sometimes. But leopards are persistent, and may stay in the tree for hours trying to reach them or waiting for one to fall in its panic, and they rarely leave until they have been successful.

Vervet Monkey (*Cercopithecus pygerythrus*)

The vervet monkey is one of the group of savannah guenons, the common long-tailed and slender-limbed monkeys of the great arc of African grasslands and wooded savannah surrounding the equatorial rain forest from the Atlantic coast to the Red Sea and down to the Cape of Good Hope. The other species are the green monkey (*C. sabaeus*) of West Africa, and the grivet monkey (*C. aethiops*) of Sudan, Eritrea, and Ethiopia, from where the range of the vervet monkey continues down to South Africa. It is a small, stocky animal, olive-green in color with white underparts and gray limbs, and has a distinctive fringe of white hair surrounding its black face.

Despite its grassland habitat the vervet monkey is primarily arboreal, and unlike the patas monkey, which leaves the trees to flee, it rarely ventures far from trees and rushes back when alarmed. Its preference is for riverine forests, which are the narrow strips of well-treed vegetation bordering the savannah rivers. It is omnivorous, its diet composed of leaves, flowers, seed pods, fruit, and grass, plus locusts, nestling birds, eggs, and small lizards. A totally diurnal species that lives in troops of up to fifty animals, it seeks the safety of trees when alarmed and sleeps in trees at night, usually in the large adjacent trees of the riverine forests, and with its light weight can get onto thin branches that will not support the leopard. Vervet monkeys may therefore be more secure from attack at night, but they are not safe from leopards that hunt by day, and lie in ambush for them. They are also taken by the caracal, serval, and African golden cat, and they never seem to learn not to trust baboons, which occasionally snatch and eat their babies.

Notes

1. In South America large rodents (the capybara and paca) are major prey species of the large cats—the jaguar and cougar.

2. This type of foot is termed paraxonic.

3. This type of foot is termed mesaxonic.

4. The other species are the Central American Baird's tapir and the woolly or mountain tapir of the northern Andes.

5. Except the babirusa of Sulawesi which is quite herbivorous, and has the beginnings of a compartmentalized stomach for the breakdown of fiber.

6. The other species are the white-lipped peccary and the larger Chacoan peccary.

7. Six races or subspecies of the elk are currently recognized. They are the eastern, Tule, Roosevelt, Rocky Mountain, Merriam's, and Manitoba Elk.

8. Grizzly bears, which are diurnal hunters, can run down young elk and are a major predator in Yellowstone National Park.

11 Primitive Primates

The primates began their evolution from ancient insectivorous mammals in the late Eocene Epoch about 55 million years ago. The major changes involved were the development of overlapping or stereoscopic vision, improved daylight and color vision, a larger and more complex brain, and the development of a human-like hand with good opposability and dexterity and finger pads with nails instead of claws. These improvements happened slowly throughout the primates' evolution from the first primitive species, which are known as prosimians, to the more advanced anthropoid[1] primates that diverged from them. However, the development of the anthropoids was not totally at the expense of the primitive prosimians and several were so successful they have survived to this day, and show a combination of their ancestors' anatomy, senses and ability, with aspects that are decidedly more advanced. They have large eyes and binocular vision, in its beginnings in some and well developed in others—especially the long-leaping bushbabies and tarsiers, which need good stereoscopic sight for depth perception. A tapetum lucidum, which reflects light back through the retina to improve nocturnal sight, is present in all species except the tarsier.

Despite their improved sight, however, the sense of smell is still important to the prosimians, and the nocturnal species still depend to a large extent upon this sense, except for the tarsier, which relies largely upon its excellent vision. The prosimians have the naked, moist nose of the insectivores, long and pointed muzzles to accommodate more membranes to detect scent, and a well-developed olfactory lobe of the brain to interpret smells. Their brains, while larger than those of their ancestral insectivores, are smaller relative to their size than those of the more advanced primates. Their hands have good grasping ability, yet they still have claws on some digits. Whereas all the anthropoids except one are diurnal and have color vision, the prosimians are mostly nocturnal and cannot perceive color. Most species have large outer ears for sound collection, and either an acute or good sense of hearing.

There are 233 species of primates, but only thirty-five are nocturnal, and with one exception they are all prosimians. They are the lemurs, bushbabies, lorises, pottos, and tarsiers, and unlike the diurnal monkeys and apes, which include many ground-dwelling species, the nocturnal primates are all arboreal and seldom leave the safety of the trees. They are animals of the warmer regions of the world. The forests of Madagascar are the most well represented with sixteen nocturnal species, and there are eleven in Africa, seven in the vast forested stretch of southern Asia from India to Indonesia and north to the Philippines, and just a single species in the neotropics—the douroucouli—which is the only nocturnal anthropoid. Howler monkeys, which are often said to be nocturnal, are actually active during the day, although they may roar early in the morning before they get up and at dusk when they are settling down for the night.

Some of the Species

Lorises and Pottos

The lorises and pottos are a group of small and very non-monkey-like primates, members of the family *Lorisidae* and collectively known as lorisoids. There are five species, the potto and angwantibo are African, and the three lorises hail from southern Asia and Indonesia. They have large eyes, small ears, a small, pointed muzzle and their coat is short, woolly, and quite dense, and excluding the potto they are tail-less. The lorisoids are all nocturnal and arboreal and normally shelter in tree holes during the day. As they are adapted for climbing their long legs are of similar length without the muscular development of the hindlimbs seen in the leapers. Also, their strong feet and toes have a wide angle (180 degrees) of opposability between the first and second digits (or the third digit in some as the index finger is quite small), and they have a very tight grip. Their wrist and ankle joints are very flexible and they have unusual blood storage channels in their hands and feet, which allows them to grasp tightly for long periods without suffering muscle fatigue.

Slender Loris (*Loris tardigradus*)

With its slender build and long, thin legs the slender loris is the scrawniest of all the primates, but it has a coat of thick and woolly fur. Its eyes are its most prominent feature; they are very large and are set close together, and their overlapping visual field results in increased powers of stereoscopic sight, although it still relies on its sense of smell when feeding. The slender loris is tail-less and when adult weighs about 12 ounces (340 g); it has a yellowish-brown coat with silvery-buff underparts, a pale face with dark circles around the eyes, accentuated by the white nasal ridge. It is a solitary animal of southern India and Sri Lanka, where it is a totally nocturnal forest-dweller. Insects, especially noxious caterpillars and ants, form the bulk of its diet, plus fruit and flowers, snails, and birds' eggs and nestlings. It is a stalker that creeps up silently to an insect, leans forward to sniff its victim, and then quickly seizes it with both hands. It is a less ponderous animal than the related slow loris, and can move fairly quickly when necessary, especially on the ground, although it rarely leaves the trees. Slender lorises sleep during the day in a tree hollow or curled up on a branch with their heads tucked between their legs, their opposable fingers and toes enabling them to grasp branches very firmly. The loris's single baby clings tightly to its mother for the first few weeks but is then left on a branch when she goes hunting.

Slow Loris (*Nycticebus coucang*)

The slow loris is very similar to the slender loris, but has a heavier build and is a much larger animal, reaching a weight of 4 pounds 6 ounces (2 kg). It also has a wider range, occurring from Bangladesh to the Malay Peninsula, Indonesia, and the Philippines. It has the same short and woolly fur, but is light reddish-brown above and buffy-gray below; it has a pale face with dark eye rings, and is tail-less. In typical loris fashion it is usually a slow and methodical mover, approaching its prey so quietly and slowly that it does not suspect danger, then moving fast to seize it with both hands while anchored to a branch with its feet. Its thumb is perpendicular to its fingers, and it is believed to have the strongest grasp of all animals in relation to its size, grasping branches so tightly it is difficult to unlock their grip, and when moving along a branch it seldom releases more than two limbs at the same time. The slow loris is nocturnal and arboreal, and is apparently less solitary in the wild than the slender loris. It is mainly insectivorous but also eats nestling birds, eggs, small lizards, and ripe fruit. It has excellent night vision and its pupils become vertically elliptical in contraction in bright light.

Potto (*Perodicticus potto*)

A single species within its genus, the potto is one of the two African primates[2] that closely resemble the Asiatic lorises. It lives in the forests of west and central Africa, and is very similar in size and shape to the slow loris, but is distinguished by its short tail. The potto weighs up to 3¼ pounds (1.5 kg), its tail averages 3 inches (8 cm) long, and it has a dense and woolly reddish-brown coat. It has a rudimentary index finger and large digit pads; its large toes and thumbs are opposable, providing it with a powerful grip, with which it can cling so firmly to a branch that it is difficult to dislodge without causing it harm. The potto has spinal processes (extensions of the backbone) which penetrate the skin of the nape and back, and when attacked it grasps a branch with both hands and delivers a violent blow with its bony spine. Like the other lorisoids it is omnivorous, preferring insects but also eating fruit and berries, birds' eggs and nestlings, and small arboreal lizards and possibly tree frogs.

Slow Loris   *The slow loris is a strictly nocturnal prosimian or primitive primate from the forests of Southeast Asia and Indonesia. Slow and methodical until it gets within reach of an insect, it is aided in its nighttime activities by acute vision and the powerful grip afforded by its opposable big toes and thumbs.*
Photo: David Haring, Duke University

The potto is totally nocturnal and sleeps by day in foliage or in a tree hole. Despite its large eyes and excellent night vision, the potto moves slowly along horizontal branches, locating its food by smell, but its binocular vision obviously assists in focusing upon and then seizing its prey. Initially the baby potto clings to its mother, but from the age of three weeks it is left clinging tightly to a branch while she goes foraging. They both spend the daylight hours in a nest, which is usually in a tree hollow.

Bushbabies

The family *Lorisidae* is also represented in Africa by nine species of endemic nocturnal primates known as bushbabies or galagos, which are totally different in both appearance and behavior from the tail-less, ponderous lorises and pottos. Bushbabies are active, agile, and very highly strung animals, capable of prodigious leaps, over 20 feet (6 m) in length and up to 7 feet (2.1 m) high in a standing jump, achieved with their strong thigh muscles and elongated hindlimbs, long tails for balance, and thickened skin pads on their digits for a good grip. Their grip is not as strong as the loris's, however, as their digits are not as completely opposable. They would not be able to make such leaps without the benefit of binocular or stereoscopic vision.

Bushbabies have slender bodies, long and very bushy tails, and thick and woolly coats. Whereas the pottos and lorises are mainly solitary animals, the bushbabies have a social structure—an extended family group arrangement—similar to the marmosets, and they have a wide vocabulary for communicating with each other. The smaller species are gentle, docile animals, which are often kept as pets in their native lands. Bushbabies are purely African animals, from the forests, wooded grasslands, and riverine woodland south of the Sahara. They have very large eyes, especially the Senegal bushbaby and the dwarf bushbaby, and good night vision. All species have large and mobile naked ears, which they can furl, and their hearing is acute. They have a wide vocabulary for communicating with others in their group, clicking and "barking," and have a high-pitched alarm call. They also have a good sense of smell, and the needle-clawed bushbaby, which feeds mainly on tree gums and may visit several hundred sites per night, locates these by smell. Bushbabies also comunicate through smell, by urinating directly onto branches, and by "washing" their hands and feet with urine, to mark their route through the trees.

Senegal or Lesser Bushbaby (*Galago senegalensis*)

The Senegal bushbaby is one of the smaller and most widespread species of bushbaby, a totally arboreal animal that usually lives in family groups, in forest, bushland, and acacia savannahs throughout sub-Saharan Africa. With their large eyes and ears and thick, fluffy coat these bushbabies are very appealing animals, and were a favorite pet species for many years prior to the recent trade bans. Several races are recognized, the most familiar being the Moholi bushbaby (*G.s.moholi*) of South Africa, whose coat resembles a chinchilla's. These small primates are similar in size to marmosets, weighing up to 10 ounces (283 g), but they have a long, thickly furred tail and their coat is thinner and shorter than the greater bushbaby's. The Senegal bushbaby is a very agile animal, capable of making tremendous leaps, its fingers and toes having flattened disks of thick skin to improve its grasp on branches. Highly nocturnal, at night its deep topaz eyes gleam intensely red when caught in a beam of light. During the day it sleeps in tree holes, usually with several others, or in an old bird's nest in dense vegetation, often close to the ground. Trappers catch them easily as they are so drowsy and dazed by bright light when awakened. Their Asian counterparts are the tarsiers and like them they are highly insectivorous, and subsist mainly on locusts and grasshoppers.

Greater or Thick-tailed Bushbaby (*Galago crassicaudatus*)

This is the largest bushbaby, similar in size to a small domestic cat with a thick and woolly buffy-gray coat with darker underparts; plus large bare ears and a thickly furred tail about 18 inches (46 cm) long. Melanistic individuals frequently occur in this species. The greater bushbaby lives in family groups—of an adult pair with their latest young—in central and southern Africa, and is highly nocturnal and arboreal, occurring wherever there are trees, even in urban areas. It does not rely on leaping as much as the other bushbabies, preferring to walk or run along horizontal branches, and to creep up stealthily on its victims like the lorises. It does not make nests in the wild but the family sleeps cuddled together in dense foliage or in a tree hole, with their heads curled down between their legs. On the rare occasions that they venture to the ground they bound along like kangaroos, holding their tails stiffly behind them as balancing organs. This species is the most carnivorous of the bushbabies, its diet comprising birds and their eggs and nestlings, lizards, and small rodents, and it has killed roosting birds as large as the chicken-sized guineafowl and spurfowl, merely to eat their brains. When kept as pets in South Africa some have become quite vicious as they matured.

Lemurs

The lemurs are monkey-like animals whose ancestors once occupied much of the world and were especially plentiful in the Northern Hemisphere. For millions of years now they have been confined to the islands of Madagascar and the Comoros in the southern Indian Ocean, where they evolved in the absence of the higher primates developing elsewhere. It is unclear when they actually arrived in Madagascar, for the island separated from Africa about 120 million years ago, but ocean-level fluctuations may have created a land link between them across the Mozambique Channel about 40 million years ago, allowing colonization then by the prosimian primates. However, lacking competitors and with many available niches and fewer predators than on the mainland, the lemurs evolved into a variety of species ranging from the tiny mouse lemurs to the indri, the largest living prosimian, which may weigh 20 pounds (9 kg). The truly nocturnal lemurs are mostly tiny animals; they include the seven species of mouse and dwarf lemurs, which are the smallest species; seven species of sportive lemurs, which all weigh about 2 pounds (900 g) when adult; the slightly larger Avahi; and the largest nocturnal species, the aye aye, which reaches a weight of about 4 pounds (1.8 kg). The remaining species of lemurs are all larger animals that are generally quite diurnal in their activity, although the gentle lemur, ruffed lemur, and the brown lemur all have varying periods of activity, are occasionally active at dusk and dawn, and could therefore be considered partially crepuscular.

Lesser Mouse Lemur (*Microcebus murinus*)

This tiny animal weighs just 1½ ounces (42 g), and has a head and body length of 4½ inches (11.5 cm), just slightly larger than the pygmy mouse lemur (*M. myoxinus*), which is the world's smallest primate. It has short, buffy or grayish-brown fur, with paler underparts and a pale stripe from the nose to above the eyes, and is often mistaken for a tiny bushbaby. A very agile leaper, it uses its tail for balance, and is assisted in its active lifestyle by its large eyes, which face forward for good stereoscopic (3D) vision. Its thin and bare membranous ears provide good sound reception. The lesser mouse lemur does not appear to have a very effective thermoregulatory mechanism, and captive animals became semitorpid and did not reproduce when kept at temperatures below 68°F (20°C). The lesser mouse lemur lives in the coastal forests of southern and western Madagascar, where it favors the lower secondary growth vegetation. It is a highly nocturnal animal; it lives in harem groups of a male and several females, and communicates with a variety of high-pitched calls. During the day they sleep in a nest of leaves or in tree hollows. In the wet season, they store fat in their limbs and tail, almost doubling their normal body weight, and then become torpid during the dry and cool winter season, when several huddle together in a tree hollow and are quite cool and stiff to the touch.

Fat-tailed Dwarf Lemur (*Cheirogaleus medius*)

This is another small species of dormouse-like lemur, from the seasonally dry coastal forests of western and southern Madagascar. It is slightly larger than the lesser mouse lemur, weighing about 5 ounces (140 g), and has soft and woolly fur, grayish-brown above and yellowish-white below, a furred tail, and large eyes for nocturnal activity. A highly nocturnal forest-dweller, it spends the daylight hours in a spherical nest of leafy twigs in the treetops or in a tree hollow, and appears at dusk to search for insects, fruit, and plant sap. It stores fat in its tail, almost doubling its weight, for the period of winter hibernation, and may also estivate during periods of intense heat, usually passing these periods of torpidity in the relative security of a tree cavity. Despite the storage of fat for estivation, it still loses about one quarter of its body weight during these long periods of sleep. The fat-tailed lemur has always been considered a solitary species, but colonial behavior has been reported in wild individuals, which were discovered estivating together. Basically a very quiet animal, it makes low sounds when communicating with conspecifics and louder calls when alarmed. After a gestation period of just sixty-one days, three or four babies are born in the tree nest, fully furred and with their eyes open. Like all Madagascar's lemurs it is declining due to forest clearance for agriculture, and it must still cope with its main predators, which are owls and the fossa (*Cryptoprocta ferox*), a very agile relative of the mongoose.

Aye Aye (*Daubentonia madagascariensis*)

The aye aye is the largest nocturnal prosimian and nocturnal primate, adults weighing about 4 pounds (1.8 kg), and is a most unusual animal with a round head and short face, large orange eyes, large naked ears, and a pointed muzzle. It has a slender body and long limbs, and a coat of long, coarse black hair. Its senses are well developed and it has good night vision, its hearing is exceptional, and it has a good sense of smell. When adult it has large, curved, rodent-like incisors and no canines, and there is a large space between the incisors and the premolars, similar to the rodent's diastema. The aye aye's fingers are also unusual, for they are long and thin; the third finger is extra long and is used for feeding and personal grooming—for scratching, combing, and cleaning the hair. The thumb is not fully opposable, unlike the first toe, and all the digits have claws except the first toe, which has a flat toenail. Insects and fruit form the bulk of its diet, and with its keen sense of hearing it locates beetle larvae in decaying tree trunks, bites the wood away, and hooks the larva out with its long third finger. It also eats tree exudates and picks raw coconut meat out with its special finger after first biting through the husk. The aye aye is a totally nocturnal animal of the forests and mangroves, which sleeps during the day in a woven nest of leafy branches. It is now very rare due to the loss of forest habitat, especially decaying trees infested with beetle larvae, and because of persecution by superstitious people who believe it presages evil.

Tarsiers

Philippine Tarsier (*Tarsius syrichta*)

The tarsiers are tiny primitive primates, their huge eyes, large ears, short neck, and protruding muzzle giving them an almost extraterrestrial appearance. The Philippine tarsier is one of five species of tarsiers in Southeast Asia, and lives on several islands in the Philippine Archipelago. It has short, grayish-brown fur and weighs only a few ounces; its naked tail is longer than the head and body, which measures only 6 inches (15 cm). Tarsiers are nocturnal forest-dwellers and in a typical resting position cling upright to a vertical trunk or branch. They have short forelimbs, and very long and powerful hindlimbs allowing them to leap up to 18 feet (5.5 m) between tree trunks, on which the round pads on their digits provide a good grip. Their first toe is opposable and they have extra long nails on the second and third fingers for grooming. Tarsiers are the only truly insectivorous primates, and in their search for prey they can rotate their heads 360 degrees. They have the largest eyes of any animal in relation to body size, the diameter of their eyeballs measuring 16 mm, and each eye is larger than their brain. They are so greatly

Philippine Tarsier *Totally nocturnal, and the only truly insectivorous primate, the tarsier has the largest eyes of all mammals in relation to its size, each one being larger than its brain. They are fixed and immovable and the tarsier compensates by rotating its head 360 degrees to search for its prey.*
Photo: Erik Degraaf, Shutterstock.com

enlarged, and surrounded by a bony rim, that they protrude well beyond the orbit, and are immovable, so the tarsier must move its head to change its field of vision. When it sees an insect it noticeably moves its head to bring its prey into focus and then leaps to seize it with both hands. Despite these adaptations for improved vision, tarsiers lack a tapetum lucidum, which in other nocturnal species reflects light back through the photoreceptor cells of the retina to enhance the available light. Tarsiers live in pairs with their latest young, which sometimes stay with their parents until they are mature. They are highly territorial and mark their area with urine.

Douroucoulis

Douroucouli or Night Monkey (*Aotus trivirgatus*)

A member of the *Cebidae* family, to which all the South American monkeys except the marmosets and tamarins belong, there is disagreement over the classification of the dourouroculi. Traditionally, it has been considered a monotypic species in the genus *Aotus*—with just one species and many races or subspecies. However, some taxonomists recently elevated several races to specific status, and claim that there are actually up to nine species of douroucouli. The douroucouli is a small primate of the neotropical rain forest, comparable to the squirrel monkey in size, weighing up to 2 pounds (900 g) when adult and with a long tail that equals its head and body length of 15 inches (38 cm). It has a small head and large eyes with large, white eyebrows, long limbs, and a thick brownish-gray coat with a green tinge and yellowish-gray underparts. It is the only truly nocturnal South American primate, and the only nocturnal anthropoid primate, which rests by day in tree hollows and in clumps of dense undergrowth. The douroucouli's very sensitive night vision is made possible by its enormous eyes, protuberant corneas, and a large concentration of rod photoreceptor cells in the retina for improved night vision. In contrast, the other members of the *Cebidae* family (spider monkeys, capuchins, and squirrel monkeys) have a typical diurnal monkey retina with mostly cones for daytime and color vision.

Notes

1. The monkeys, apes, and man.
2. The other is the angwantibo or golden potto (*Arctocebus calabarensis*).

Glossary

Aberrant
Animals that deviate in important characteristics from their nearest relatives.

Accommodation
The automatic adjustment of the focal length of the lens to allow the eye to focus on objects at varying distances.

Aqueous humor
Clear fluid in the eye (between the cornea and the lens), which provides nutrients and determines intraocular pressure.

Auditory meatus
The external opening to the ear canal.

Auricular
The region of the ear, the area around the ear opening.

Avascular
The absence of blood vessels, tissue that lacks a blood supply.

Binocular vision
Vision as a result of both eyes working together, blending the separate images seen by each eye into a composite one.

Biosynthesis
The synthesis (production) of chemical compounds by living cells, an essential part of anabolism—the constructive phase of metabolism where complex molecules are built up.

Blastocyst
A sphere of cells that form the pre-implantation embryo in cases of mammalian delayed implantation.

Brachydont
A low-crowned tooth in which the width of the tooth exceeds the height of the crown.

Carnivores
Members of the order *Carnivora*, which includes the cats, dogs, and hyenas. However, it is also used to denote any carnivorous animal that eats animal protein such as fish, crustaceans like crabs and krill, and the meat of mammals, reptiles, and amphibians.

Cochlea
The snail-shaped, fluid-filled cavity in the inner ear that converts sound vibrations from the middle ear into nerve impulses that travel to the brain.

Cones
Photoreceptor cells in the retina which are responsible for color vision in bright light.

Crepuscular
Active at dusk and dawn.

Diastema
A gap or space between teeth, as in the rodents.

Digitigrade
Walking on the toes, as in the dogs and cats.

Diurnal
Active during daylight.

Echolocation
Orientation by means of the echos returning from high-pitched sounds sent out by bats and cavebirds to avoid objects, and by bats to locate their prey.

Estivation
Long-term summer sleep to avoid hot and dry weather and to conserve water and energy, while surviving upon body fat.

Eustachian tube
The tube that connects the middle ear and the throat.

Family
Zoologically, a subgroup of an order. For example, the order *Carnivora* contains mammal families such as *Felidae* (cats), *Canidae* (dogs), and *Mustelidae* (weasels).

Fovea centralis
The small but sensitive area of the retina that provides the clearest vision and greatest focus. The blood vesels and nerve fibers go around it so that nothing interrupts the light's path to its photo-receptors.

Hibernation
Any form of long-term torpidity in ectotherms and endotherms to escape unfavorable winter conditions (of climate and food supply) and to conserve energy, while surviving upon body fat or external food stores.

Holarctic Region
A zoogeographic region encompassing the temperate regions of the Northern Hemisphere.

Hydrolize
To make a compound react with water.

Hyoid
A U-shaped bone between the mandible and the larynx at the base of the tongue, which supports the muscles of the tongue. In the cats that can roar (the lion, tiger, leopard, and jaguar) the hyoid bone is in two pieces joined by an elastic ligament. In the other cats it is a single rigid bone and they cannot roar.

Hypsodont
Teeth with high crowns that are usually rootless and continually growing.

Infrasonic
A frequency of sound below the range audible to humans.

Iris
The pigmented light-resistant barrier that regulates the amount of light entering the eye by opening and closing the pupil. Red-eye in flash photography is the result of the iris being unable to close quickly enough, thereby illuminating the blood-rich retina. Plural irides.

Jacobsen's organ
An extrasensory organ in the roof of the mouth in many animals, which receives scent particles carried by the air or on the tongue and transmits the information to the brain for action. Also known as the vomeronasal organ.

Lateral line
A sense organ in fish and some amphibians, visible in fish as faint lines along the sides of their bodies from gills to tail, which detects movement in the surrounding water. The sense receptors within the line are called neuromasts.

Macula
The small, specialized central area of the retina, responsible for acute central vision.

Marsupials
Primitive implacental mammals in which the young are born after a very short gestation period and complete their development in the mother's pouch or marsupium; for example, opossums, kanga-roos, and the koala.

Mesozoic Era
The "Age of Reptiles" from 248 to 65 million years ago (mya) which is divided into three time periods (Triassic 245–208 mya, Jurassic 208–146 mya, and Cretaceous 146–65 mya). The dinosaurs were its most familiar animals, which began evolving in the Triassic but were not too diverse until the Jurassic and then had vanished by the end of the Cretaceous. Birds began to evolve in the Jurassic, and the mammals also first appeared on Earth early in the Jurassic, but did not become dominant until the dinosaurs had gone. The Mesozoic Era was followed by the Cenozoic Era—The "Age of Mammals."

Nasopharynx
The area of the upper throat behind the nose.

Nearctic Region
The zoogeographic region encompassing the northern temperate zone of the New World.

Neocortex
The outermost convoluted area of the forebrain, often just called the cortex.

Neoteny
The retention of larval characters into adulthood. The amphibian larval form (tadpole in frogs and toads) is ideally adapted for an aquatic existence, and species that elected to stay in water as adults therefore retained this form. The axolotl is the classic example of this phenomenon, retaining its gills and long tail as an adult.

Neuromasts
Groups of hair cells that are the sensory receptors in the lateral line organs of fish and some amphibians.

Nictitating membrane
The "third eyelid" or semitransparent membrane of birds, reptiles, and amphibians, which can be drawn across the eye to protect it from light and dust.

Nocturnal
Active during the hours of darkness.

Olfactory
Relating to or connected with the sense of smell.

Opsins
Molecules in cone cells that bind to pigments and create a complex sensitive to light of certain wavelengths.

Otic
Relating to or near the ear.

Palaearctic Region
The zoogeographic region encompassing the northern temperate zone of the Old World.

Patagium
The expandable membrane connecting the limbs of certain mammals, which permits either full flight (by bats) or gliding (by the flying squirrels and sugar gliders).

Pecten
A vascular structure projecting from the retina into the vitreous humor, from which nutrients and oxygen diffuse.

Pheromones
Chemical signals that travel between organisms as a form of communication to trigger a social or sexual response.

Photoreceptors
Pigment-filled sensors (either rods or cones) at the back of the retina which are sensitive to light.

Plantigrade
Walking on the entire sole of the foot, as in the bears and humans.

Proprioceptors
Sensory nerve endings in the muscles, tendons, and joints, which provide a sense of position by responding to stimuli from within the body.

Pupil
The hole in the center of the iris which appears black because light entering it is absorbed by the tissues of the eye. Pupil size determines the amount of light entering the eye, and is controlled by involuntary contractions and dilations of the sphincter and dilator muscles of the iris.

Receptors
Specialized structures on the surface of cells, which allow them to respond to hormones, neuro-transmitters, and other message-bearing molecules.

Retina
The light-sensitive tissue at the back of the eye that transmits visual impulses to the brian via the optic nerve.

Rhinarium
A naked moist patch of skin surrounding the nostrils, which is especially large in pigs.

Rhodopsin
A pigment, also known as visual purple, found in the external segments of the retina, that is sensitive to red light.

Rods
Photoreceptor cells that respond to dim light and are responsible for peripheral and night vision.

Scutes
Large scales of horny keratin on the outer layer of turtles' shells.

Sensory organ
An organ capable of receiving and responding to outside stimuli, such as the lateral line organs of fish and some amphibians.

Somatic
Relating to the nonsexual tissues and organs of the body.

Stereoscopic
Three-dimensional vision that allows good perception of depth.

Tapetum
A reflecting layer behind the retina that sends light back to the retina, increasing the quantity of light and thus improving vision in the dark. Responsible for the "eye-shine" of animals caught in the headlights.

Trichromatic
Full color vision, involving red, green, and blue to create the range of colors in the spectrum.

Turbinates
The structures inside the nose that humidify and filter air.

Tympanum
The eardrum, a membrane separating the external ear from the middle ear chamber, and visible in most frogs on the sides of their heads.

Ultrasonic
A sound frequency too high for the human ear to hear.

Vestigial
Small and no longer useful structures that were developed and functional in the animal's ancestors, but have become rudimentary during the course of evolution.

Visceral
Relating to the main soft internal organs such as the heart, lungs, and digestive system.

Vitreous humor
The transparent jelly that fills the eyes between the lens and the retina, and maintains the shape of the eyeball.

Vomeronasal organ—see Jacobsen's organ

Bibliography

Alterman, L. and Doyle, G. A. *Creatures of the Dark: The Nocturnal Prosimians*. Plenum Press, New York, 1995.

Animal Facts. http://www.bbc.co.uk/nature/wildfacts/.

Bats. http://www.batcon.org/.

Biology. Ask the Experts. http://www.sciam.com/askexpert_question.

Bogert, C. M. How reptiles regulate their body temperature. *Scientific American* 200 (1959): 105.

Busse, C. Leopard and Lion predation upon Chacma Baboons living in Moremi Wildlife Reserve. *Botswana Notes and Records, Biol.* 12 (1980): 15–21.

Campbell, B. and Lack, E. (Eds.). *A Dictionary of Birds*. Buteo, Vermilion, 1985.

Church, D. C. *Digestive Physiology and Nutrition of Ruminants*, Vol. II. O and B Books, Corvallis, OR, 1979.

Conant, R. *A Field Guide to Reptiles and Amphibians of Eastern/Central North America*. Houghton Mifflin, Boston, 1975.

Crowcroft, P. *The Life of the Shrew*. Reinhardt, London, 1957.

Dictionary. MSN Encarta. http://ca.encarta.msn.com/encyclopedia.

Dryden, G. L. Food of wild and captive insectivora. In *Handbook Series in Nutrition and Food*. Vol. I. Edited by M. Rechcigl. CRC Press, Cleveland, 1977, 469–492.

Eisenberg, J. F. The Heteromyid Rodents. In *The UFAW Handbook on the Care and Management of Laboratory Animals*. Livingstone, Edinburgh and London, 1966, 391–395.

Endangered Species. ARKIVE. www.arkive.org/species.

Gould, E. and Eisenberg, J. Notes on the biology of the *Tenrecidae*. *J. Mamm.* 47 (1966): 660–686.

Griffin, D. R. *Listening in the Dark*. Yale University Press, New Haven, 1958.

Hoser, R. T. *Australian Reptiles and Frogs*. Pierson & Co., Mosman, NSW, 1989.

House, H. B. and Doherty, J. G. The World of Darkness at the New York Zoological Park. *International Zoo Yearbook* 15 (1975): 51–54.

Inger, R. F. and Stuebing, R. B. *Frogs of Sabah*. Sabah Parks Pubn. No. 10, Kota Kinabalu, 1989.

Keng, F.L.L. and Tat-mong, M. L. *Fascinating Snakes of Southeast Asia—An Introduction*. Tropical Press, Kuala Lumpur, 1990.

Kruuk, H. and Turner, M. Comparative notes on predation by lion, leopard, cheetah and wild dogs in the Serengeti. *Mammalia* 31 (1967): 1–27.

Landry, S. O. The rodentia as omnivores. *Q. Rev. Biol.* 45 (1970): 351–372.

Milne, L. and Milne, M. *The senses of animals and men.* Athenaeum, New York, 1962.

Nocturnal Desert Animals. http://www.pbs.org/nova/kalahari/.

Nocturnal Eye, The. 2000. Nova On Line. www.pbs.org/wgbh/nova/kalahari/nightvision.html.

Nowak, R. M. *Walker's Mammals of the World.* Vols. I & II. Johns Hopkins University Press, Baltimore and London, 1991.

Prater, S. H. *The Book of Indian Mammals.* Bombay Natural History Society, Bombay, 1965.

Primates. http://www.Animalomnibus.com/primates.htm.

Roberts, A. *The Mammals of South Africa.* The Mammals of S. Africa Book Fund, Johannesburg, 1951.

Roots, C. *Animals of the Dark.* Praeger, New York, 1974.

Russian Bats. http://www.zmmu.msu.ru/bats.rbgrhp/autjor1.html.

Species information and classification. ADW. http://animaldiversity.ummz.umich.edu/site.

Stebbins, R. C. *A Field Guide to Western Reptiles and Amphibians.* Houghton Mifflin, Boston, 1966.

Tattersal, I. *The Primates of Madagascar.* Colorado University Press, New York, 1982.

Whitehead, G. K. *Deer of the World.* Viking, New York, 1972.

Wild Carnivores. Lioncrusher's Domain. http://www.lioncrusher.com.

Index

About the Author

CLIVE ROOTS has been a zoo director for many years. He has traveled the world collecting live animals for zoo conservation programs. Roots has acted as a masterplanning and design consultant for numerous zoological gardens and related projects around the world, and has written many books on zoo and natural history subjects.